日本茶

JAPANESE GREEN TEA

东京艺术之旅

TOKYO ARTRIP

〔日〕美术出版社书籍编辑部　编著

黄迪　译

中信出版集团 | 北京

前言

旅行指南系列《东京艺术之旅》，每册一个主题，带你从日本文化、艺术及设计的角度畅游东京。每册均有不同的艺术之旅顾问登场，本册以日本茶为主题。三位专家与一家正宗的日本茶铺将为大家带来详尽的“东京日本茶指南”。日本茶讲师兼二级建筑士的柳本茜将在第一章为大家担任品茶顾问，茶艺师和多田喜则会在第二章为大家讲解日本茶的种类，日本茶爱好者的圣地——一保堂茶铺东京丸之内店会在第三章带我们领略日本茶的真正魅力，第四章则是由瑞典籍的日本茶讲师奥斯卡·布莱克鲁带我们逛遍东京的日本茶好店。

Introduction

TOKYO ARTRIP is a series of guidebooks about Tokyo. Each playful, walking spot introduced in each series is selected from the perspective of art & design. 4 ARTRIP ADVISERS appear in each series. In this *Japanese green tea*, the following 4 ARTRIP ADVISERS appear in the book. Ms. Akane Yanagimoto who is a Second-Class Architect and Nihoncha Instructor (PART 1); Mr. Yoshi Watada who is a Nihoncha Sommelier (PART 2); IPPODO TEA’ s Marunouchi Store known as the holy ground for nihoncha lovers (PART 3); and Oscar Brekell who is from Sweden and a Nihoncha Instructor (PART 4). In this book, we take you to playful, nihoncha spots in Tokyo recommended by these 4 experts.

目录

符号说明 (时) 营业时间 (电) 电话 (休) 休息日 (址) 地址 (交) 交通路线 (网) 网址

※ 价格全部不含税。
※ 书中内容基于 2017 年 10 月信息。

CONTENTS

Icon Description (H) Hours of Operation (T) Telephone Number (C) Closed Days (Ad) Address (Ac) Access (U) URL

※All prices do not include sales tax, unless otherwise stated.

※All information contained in this book are as of October, 2017.

第一章

品茶：
与空间一起品味

ENJOY NIHONCHA:
ENJOY A CUP WITH
SURROUNDING SPACE

若要探求日本茶的美味，享用茶品的空间与环境很重要

“茜夜”茶店位于神乐坂的僻静之处，深夜也能为客人提供现泡的热茶。店主柳本茜的主业是建筑师，店内的装修都是她亲力亲为设计的。柔和的灯光下，摆放着天然质地的旧杂货小物与家具，使来客身心都能得到治愈。店主说：“所处环境的不同，亦能品出茶不同的滋味。”在和式的简约空间里饮茶，与在现代风格的建筑物里饮茶，味道各不相同，但都让人回味。店主认为，要想品出茶的美味，最好要有意识地选择所处的空间。

If you seek a beautiful cup of nihoncha, then be conscious about what kind of space and environment you drink your cup in.

Hidden in a neighborhood of Kagurazaka, AKANE-YA (page 18) is known to those in the know as a place where you can enjoy properly brewed nihoncha, even late at night. Ms. Akane Yanagimoto is the owner of AKANE-YA. She is a full-time architect, so she handled AKANE-YA's interior design. A space with soft atmosphere and dim lights, its antique-style décor items and furniture are rather spontaneously arranged. Visitors can enjoy nihoncha at night, while feeling at ease by AKANE-YA's interiors and its space. "The taste of nihoncha changes depending on where we drink it." Akane says. Whether you enjoy nihoncha at a dignified, elegant, traditional Japanese style setting, or in a modern building, you will enjoy your cup. "I think if you wish to drink a beautiful cup of nihoncha, it is good to be conscious about what kind of space you drink it in." she explains.

ARTRIP ADVISER
艺术之旅顾问

柳本茜
Akane Yanagimoto

平面设计师、二级建筑士、日本茶讲师、清酒侍酒师，出生于静冈县浜松市（在日本，县是比市高一级的行政区划）。是主营日本茶和酒的店铺——“茜夜”的店主，出版了《日本茶最佳的泡法》《茜夜的小确幸岁时记》等著作。

Graphic Designer, Architect, Nihoncha Instructor & Sake Sommelier. Born in Hamamatsu. Owner of AKANE-YA. Her books, *Ichiban Oishii Nihoncha no Irekata - The Best Cup of Nihoncha*, and *AKANE-YA no Chiisaku Tanoshimu Ouchi Saijiki - Enjoy Your Home in AKANE-YA Style*.

❶ 八云茶寮（目黑）

在目黑住宅区一个闲静的角落内，有一处显眼的大门，大门内被绿叶簇拥的宅邸显得颇为神秘。拾级而上，掀起白色的暖帘，八云茶寮静谧地矗立在人们的眼前。茶寮于 2009 年开业，是一幢由在昭和时代建成的个人住宅改造而成的简约建筑。内墙、地面、照明以及摆放的家具都不失个性，但整体很协调，给人以清静的感受。在店内的茶房，客人可以享用配以茶品的粥或者白饭的“早茶”。略显匆忙的早晨，唯有在这郁郁葱葱的庭院，或是在弥漫着茶香的室内，才能享受到这份闹中取静的恬淡。中午则供应由技艺精湛的厨师烹制的“怀石午餐”，下午还能选择来一份“午申茶”茶点——精选的日本茶或抹茶，搭配季节性的甜点与和果子。

㊋ 9:00—17:00（最后点单时间 16:00）㊊ 03-5731-1620 ㊡ 星期日、节假日（周一及节假日后第一天无早茶、怀石午餐供应）㊣ 目黑区八云 3-4-7 ㊋ 东急东横线都立大学站步行 15 分钟 ㊋ yakumosaryo.jp

❶ yakumo saryo (Meguro)

A mystical gate covered with greenery appears in a corner of a quiet residential area. Take stone steps up and go through the gate under its white noren, a shop curtain. You will then find yakumo saryo, stands quiet and dignified there. Built during the Showa period as a mansion, it was stylishly renovated and turned into the Japanese dining club and tea house in 2009. Interiors are individually unique yet harmonious, making the space comfortable. At sabou, meaning tea house, you can enjoy Asacha, a nihoncha course with either okayu (rice soup), or hakuhan (white rice). Nihoncha is brewed with care here, using chagama,a special tea pot. Enjoy the fragrance and flavor with your five senses. Viewing its garden with deep greenery from the room reminds you of chashitsu, a traditional tearoom. For lunch, the Japanese dining club serves Hiru-Kaiseki, a beautiful Japanese cuisines, served with tea. In the afternoon, it serves Goshincha, a set of seasonal wagashi (Japanese confections), with nihoncha and matcha.

Ⓗ 9:00-17:00 (L.O.16:00)(Only by inquiry for dinner) Ⓣ 03-5731-1620 Ⓒ Sunday and Holidays (No Asacha or Hiru-Kaiseki on Monday and the day after Holiday) (Ad) 3-4-7 Yakumo, Meguro-ku (Ac) 15-minute walk from Toritsudaigaku Station (Tokyu Toyoko Line) Ⓤ yakumosaryo.jp/e

在茶寮的深处有一间现代茶室，烧水的炉子摆在大大的木桌上，桌边整齐摆放着极具设计感的矮凳，这样的空间布置能让人联想到茶道中所提倡的“结界”一说。装饰墙面的卷轴会配合沙龙活动主题以及季节有所调整。

Sabou, in the back of the mansion, is a modern chashitsu. Grid patterned chairs surrounding a huge table with ro, a traditional Japanese hearth, and kama, Japanese tea pot, remind visitors of kekkai, a boundary marker used during the traditional tea ceremony. Kakejiku (Japanese hanging scrolls), and other decorations change each season as per the seasonal exhibition held at the tearoom.

“早茶”的第一道茶会搭配季节性的食材和水果，玻璃材质的透明茶具使人可以清晰地看到茶品的色泽，让人不禁期待接下来的菜品。

The first tea for Asacha is paired with seasonal ingredients and fruits. Tea is served in a clear kyusu and a cup, so you can enjoy the color and the design. It builds anticipation for the Asacha course items to follow.

“早茶”套餐需要提前预约，价格为3500日元。套餐内含季节性新茶、现炒焙茶、白粥或米饭、当季食材、鱼干、小菜、腌菜小碟、味噌汤，以及生果子和薄茶。

Asacha (3500 yen) by reservation only. Asacha course starts with seasonal tea, and includes homemade Hojicha, okayu or hakuhan with a seasonal plate, a dried fish, meshi no tomo (a small dish goes well with rice), and ko no mono (a dish of pickled vegetables), miso soup and is finished off with a fresh dessert and ousu - whisked matcha.

生果子与薄茶不仅在“早茶”套餐能享用，中午的怀石料理套餐中也有供应。若您坐在位于茶房后方的吧台位置，可以欣赏到茶道点茶的过程。

Hiru-Kaiseki course also finishes with a fresh dessert and ousu. In the back area of sabou at the counter, you can see otemae, a proper manner in which the matcha is prepared.

大长桌的位置偏低，客人坐下后如同坐在榻榻米上的暖桌前，刚好能平视窗外。客人边欣赏窗外的景色，边享受刚泡好的热茶以及美食，别具一格。

A huge table is set a little low helping the customers see windows from the same height as if you are sitting on the tatami with a table. Nothing is better than being able to enjoy the view over a cup of carefully made tea and foods. A cup of tea is served with water boiled with a chagama and evaporating steam from chagama helps add a unique atmosphere to the experience.

店主巧妙利用了入口处规划的售卖空间“梅心果”，在这里摆放了店铺的招牌茶叶以及和果子。空间内侧的沙龙区展示售卖茶房内使用的茶具，还会根据季节举行企划展览。

Baishinka is a retail space utilizing a space by the entrance. Chaba- nihoncha tea leaves, as well as seasonal high-quality fresh sweets are lined up here. At the tea house in the back, chaki used at sabou are on display and for sale. Exhibitions are held here for each season.

在这里你可以购入季节性的蒸茶、自制的配方茶以及焙茶。包装精美，作为伴手礼颇有人气。

You can purchase chaba, such as Seasonal Sencha, Homemade Blend Nihoncha and Hojicha. Beautifully wrapped teas look pristine. It is popular for omotase, a small gift to bring when people are visiting someone.

“梅心果”店铺提供的和果子是使用和三盆（细白糖）、黑砂糖、蔬菜等天然材料加以发酵而成的，色香味俱全，形状精巧而味道醇厚。

Wagashi are made with natural ingredients such as wasanbon (a finely grained Japanese sugar), kurozato (a cane sugar) and/or vegetables. Natural sweetness, color, flavor and umami of these ingredients are amplified by their natural essence and by fermentation. They look angelic, yet they deliver a rich taste.

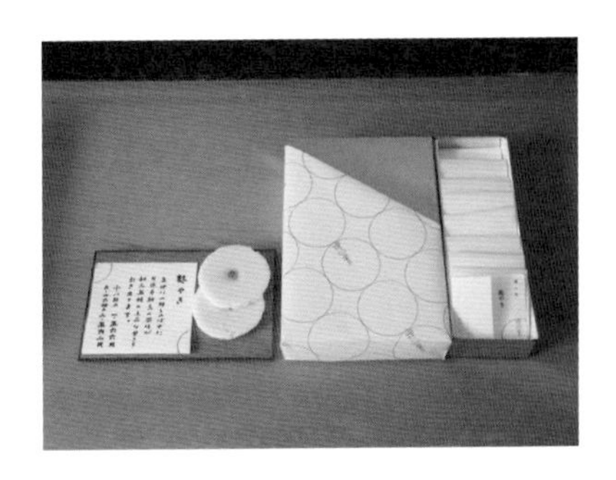

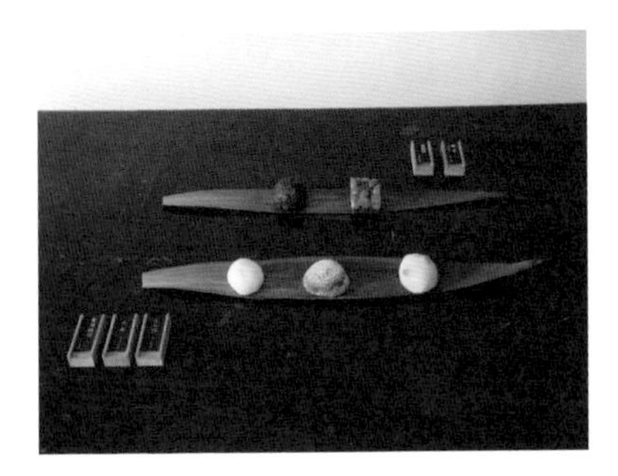

也可以购买由店内茶房供应的上生果子。精选材料，采用极致的工艺制成的点心，口感细腻。店铺提供的和果子会依季节变化更替，让人每次来店时都心怀期待。

You can also purchase high-quality, fresh sweets served at sabou. All of them are petite, elegant and each one looks like an art object. You can enjoy unique sweets made for each season, something you can keep looking forward to as a season goes by.

❷ 茜夜（饭田桥）

“茜夜”位于办公楼及饮食店林立的饭田桥，一楼是一家画廊，拾级而上，便能看到小店的招牌。正如店名中的“夜”字所指，茶店从晚上七点才开始营业。以让大众回归纸质书籍、重拾笔作记录为经营理念，店内被布置成旧时的图书馆一般，整洁而严肃，但透过偌大的窗户可以看到四季交替的不同景致，店内摆放的静物与灯光更是营造出了一种祥和的氛围。茶品主要有产自店主柳本茜的家乡的静冈茶，以及她以前居住过的福冈（八女）产的茶叶。身为日本茶讲师，她会专业细致地冲泡每一壶茶供客人享用。菜单上还有酒精饮品及季节性的食材和甜点。茜夜是一个适合各种场景的空间。到了新茶出产的时期，小店会偶尔在白天营业。

(时)19:00—23:00（最后点单时间 22:30）(电) 03-3261-7022 (休) 星期六、星期日、星期一、节假日 (址) 千代田区饭田桥 3-3-11 2 楼 (交) JR 饭田桥站东口或地铁有乐町线、大江户线、东西线饭田桥站 A2 出口步行 3 分钟 (网) www.akane-ya.net

❷ AKANE-YA (Iidabashi)

On a street corner in Iidabashi, between offices and restaurants, find a building with a gallery on the first floor. Climb the stairs, then you will find AKANE-YA. Part of the café's name, YA means night in Japanese. True to its name, AKANE-YA has unusual hours for a café, opening at 7 pm. Its concept is reading books and writing letters. It offers a library-like, inorganic, yet warm atmosphere. Through a big window, you can enjoy trees changing colors with each season. The menu is mainly composed of teas from Shizuoka, her hometown and Yame in Fukuoka, to which she feels close. Akane makes teas attentively while utilizing the techniques she acquired as a certified Nihoncha Instructor. You can also enjoy alcoholic beverages, tapas and desserts. During the nihoncha harvesting season, the café opens for special events during the daytime.

Ⓗ 19:00-23:00 (L.O.22:30) Ⓣ 03-3261-7022 Ⓒ Saturday, Sunday, Monday and Holidays (Ad) 2nd Floor, 3-3-11 Iidabashi, Chiyoda-ku (Ac) 3-minute walk from Iidabashi Station East Exit (JR Line) or Iidabashi Station Exit A2 (Tokyo Metro Yurakucho-Line, Oedo Line and Tozai Line).
Ⓤ www.akane-ya.net

1枚 1200円
ハーフ 800円

茶室原来是一个仓库，店主翻新了天花板，将空间进行了改造。从桌椅、灯具、各种小东西到茶具都是柳本店主亲自挑选的。

Akane renovated a storage space of an office by herself. She painted the ceiling on her own. All interior items, chairs, tables, lighting, petite décor items as well as tea items are carefully selected by herself.

福冈县八女市产的玉露茶（1000 日元）。将茶叶放于壶中浸泡后取出，放在小碟子里，将醇香留在杯具之中。茶壶与茶杯都是一掌可握，筷子是儿童用的尺寸，却出乎意料地易于使用。

Gyokuro from Yame, Fukuoka (1000 yen). Chaba with a deep, concentrated umami savory taste. After brewing for the tea, chaba turn into ohitashi, a dish with chaba steeped in dashi sauce. Kyusu (teapot), and chaki (a tea set), are all petite but fit into the palm of your hands. You will be eating with kids-size chopsticks, but they are surprisingly easy to use.

静冈县挂川市产的煎茶（800 日元）。客人可品尝到此款茶叶的甘甜与略微的苦涩，还可以在等待茶叶浸泡的时候，看着时间在沙漏中静静地流淌。

Sencha from Kakegawa, Shizuoka (800 yen). Enjoy a well-balanced umami, shibumi (subdued astringent taste), and nigami (bitterness). This set comes with a sand clock, so you can enjoy the passing of time while the tea is being brewed.

手工焙茶（800 日元）。店主会将静冈出产的茎茶在日式砂锅中现炒后，用开水冲泡送到客人面前。制作过程中茶香四溢，连竹制的滤茶网都会染上茶香。

Teiri - hand-roasted, Hojicha (800 yen). Kukicha is tea made from stalks and twigs of the tea plant. Green kukicha from Shizuoka, is hand-roasted in houroku, a Japanese ceramic tea roaster, on the spot. Enjoy the roasting process in front of your eyes while breathing in the toasty fragrance steaming out from the tea. Chakoshi, a tea strainer, made of bamboo, adds a tasteful design visually too.

黄金茶（1000 日元）。黄金茶是静冈县出产的一种香气较浓的煎茶，颜色金黄，可以泡上 7—8 回。店主选用透明的茶具是为了让客人能够欣赏到茶的色彩。

Ougon no Ocha, Golden Tea (1000 yen). Sencha from Shizuoka, with rare chaba and a very strong umami flavor. The color of this sencha is yellow-gold. You can add hot water 7 to 8 times for this tea. Ougon no Ocha will be served in a clear cup so you can enjoy the color of the tea.

❸ 中村藤吉本店银座店（银座）

中村藤吉本店创立于安政元年（1854 年），总店位于日本茶的起源地京都府宇治市，银座店是他们在东京的第一家分店。位于宇治市的总店被京都府指定为重要文化景观，它以古色古香的建筑风格，以及制茶工厂改造成的咖啡厅闻名；而银座分店的店内装修则是融合了和风与现代的干练风格，中村藤吉本店的屋号图案是一个“十”字外加上圆圈。银座店位于商场 GINZA SIX 内，当您掀开带有屋号图案的暖帘进入店内时，会惊讶于空间的开阔。银座店为客人准备了招牌秘制的“中村茶”，每一位进店的客人都能喝上一杯热乎乎的“中村茶”（中村茶由 7 种茶叶混合制成）。不仅如此，还有银座店限定的点心与冻糕可以选择。主食则有用茶叶与京都地方特色食材制成的菜品可选，如茶荞麦面以及抹茶乌冬等。客人可以在柜台选购中村茶等茶叶、生茶果冻、抹茶或者焙茶制成的点心。

㊐10:30—20:30（最后点单时间 19:45）㊁ 03-6264-5168 ㊡ 根据 GINZA SIX 的营业时间 ㊟ 中央区银座 6-10-1 GINZA SIX 4 楼 ㊌ 地铁银座站 A3 出口步行 2 分钟 ㊎ www.tokichi.jp

❸ NAKAMURA TOKICHI HONTEN Ginza Store (Ginza)

In 1854, NAKAMURA TOKICHI HONTEN was established in Uji, Kyoko, and has been known for its high-quality nihoncha. Designated as "the Important Cultural Landscape," the main store in Uji, was renovated from an old tea factory and offers a quaint atmosphere. Ginza store in Tokyo combines a modern design with Japanese Wa design, and welcomes its visitors with sophisticated atmosphere. Go through under the noren with the store symbol maruto sign printed on it, and you are welcomed by a spacious area. The recommended item is "Nakamuracha," special blend, made by gougumi, means blending. It is a blend of 7 kinds of chaba. Blending ratio is secret. The tea (seasonally) can be enjoyed all the time without being too concerned so much about the humidity. Nakamuracha is served as a welcome tea at the Ginza store. Some of the sweets and parfaits are only served at the Ginza store. All food items contain nihoncha or ingredients unique to Kyoto, such as Cha Soba and Matcha Udon. At the retail section, Fresh Nihoncha Jelly or other sweets are available.

Ⓗ 10:30-20:30 (L.O.19:45) Ⓣ 03-6264-5168 Ⓒ Same hours as GINZA SIX's operating hours (Ad) 10-1 Ginza 6 Chome, Chuo-ku (Ac) 2-minute walk from Ginza Station Exit A3 (Tokyo Metro Ginza Line) Ⓤ www.tokichi.jp/english

带点心的中村茶套餐（700 日元）。点单之后，会为您端上热水壶、茶壶（一般冲泡宇治茶都会使用没有把手的茶壶）、冷却热水用的容器、茶碗以及时钟。客人可以依照随餐附上的说明书自己泡茶，喝茶时可以续杯。

Nakamuracha Set with Sweets (700 yen).
The set comes with a teapot of hot water, kyusu, yusamashi (a pitcher to cool the water), yunomi (a teacup and a clock). It also comes with a how-to-brew-a-tea brochure, so you can make tea by yourself. Refills can be requested.

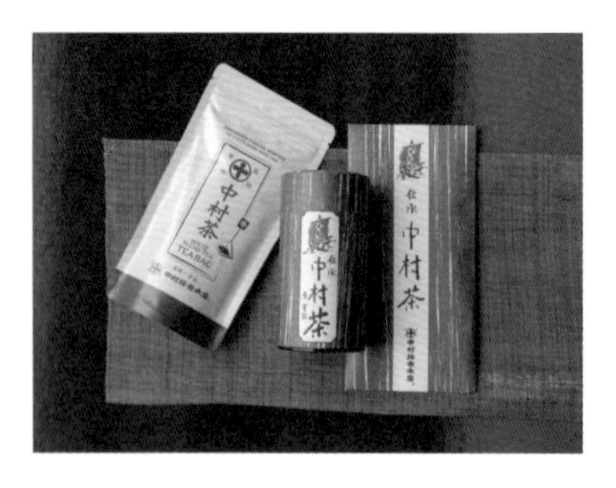

中村茶分别有 80 克的罐装（2000 日元）、50 克的袋装（1000 日元）、4 克 ×10 包的茶包装 (1100 日元)。夏天还可以买到冷泡茶套装。

Nakamuracha:
80g in a Tea Caddy (2000 yen),
50g in a Bag (1000 yen),
4g x 10 Individual Tea Bags (1100 yen).
During summer, Nakamuracha blend for a cold brew is also available.

店内的花岗岩地砖，与研磨抹茶粉的石臼是同一种材质。桌椅选择黑色，让人联想到茶商在甄选茶叶时的“审茶台”，而桌面上摆放着彩色的麻制桌布。整体搭配兼顾了京都的温润与银座的优雅气质。

The floor is made with Mikageishi, a type of granite used for Ishiusu, a stonemill used to grind matcha. The black interior design mirrors haikenba, a tea evaluation room at nihoncha retailers. The vividness of colorful hemp cloth offers beautiful contrast. Enjoy the warmth of Kyoto and the elegance of Ginza here.

❹ 东屋茶寮（银座）

东屋茶寮所在的建筑位于银座中央大道边上。客人需要乘坐电梯上至 2 层，电梯门开便进入一个欧式装修风格的宽敞空间。在设计简约的空间里即可享用日本茶与和果子，是东屋茶寮的与众不同之处。菜单上可选择季节性的水果茶——由店家严选的茶叶搭配新鲜水果或者香草制成，或者店家自制的煎茶。茶点可以选择如宝石般色彩斑斓的一口吃小点心，或者微甜的生果子。东屋茶寮是一家坚持守护传统茶道的、同时不忘研究与开发茶的喝法以及新式和果子的茶店。茶房的中央放着一个铜制的热水炉，在这里可以欣赏店员泡茶的过程。除此之外，客人还可以选择品茶的“茶食会”、日本食文化的基本之“一汤三菜”或者包含稻荷寿司与和果子的日式下午茶套餐。

(时) 工作日 11:00—22:00, 星期日、节假日营业至 19:00（打烊前 1 小时不再接单）(电) 03-3538-3240 (休) 星期一、节假日后第一天 (址) 中央区银座 1-7-7 Pola 银座大楼 2 楼 (交) 地铁有乐町线银座一丁目站 7 号出口直达 (网) www.higashiya.com/ginza

❹ HIGASHIYA GINZA (Ginza)

HIGASHIYA GINZA's elegant space will remind you of a tearoom in Europe. The tea salon offers a completely new outlook from nihoncha to wagashi and kanmi (Japanese traditional sweets). Menu includes Kisetsu no Ocha, a seasonal blend paired with fresh fruits and herbs or Original Blend Tea, paired with colorful, jewel-like one-bite kashi. While maintaining traditions, HIGASHIYA GINZA is a new, modern tearoom, offering innovative items, which you will probably experience for the first time here. Nihoncha is brewed at the center, where ro made of copper is placed. The sophisticated manner in which the staff brew the tea is also a must-see. Enjoy courses: Sajikie, a course to drink nihoncha. Ichiju-sansai, the meal consists of 1 soup and 3 dishes in traditional Japanese cuisine style. And Samajiki, an afternoon tea course with inari-zushi and wagashi.

(H) Weekday:11:00-22:00; Sunday and Holidays 11:00-19:00 (L.O.1 hour before closing) (T) 03-3538-3240 (C) Monday and the day after Holidays (Ad) Pola Ginza Building 2nd Floor, 1-7-7 Ginza, Chuo-ku (Ac) In front of Ext 7, Ginza 1 Chome Station (Tokyo Metro Yurakucho Line) (U) www.higashiya.com/en

店家有着独特的泡茶法：坐在椅子上进行点茶的日本茶道的“立礼式”，以及冲泡台湾茶时使用如左图所示的茶盘。

HIGASHIYA style of brewing nihoncha is a unique method. The manner of serving may remind visitors of ryureishiki style, a manner in a Japanese tea ceremony, which utilizes a table and chairs. Tea is served on chaban, a plate for teas, customarily used by Taiwanese Tea brewing, and this also adds to the uniqueness of the experience at HIGASHIYA.

店内设计以原木色为基调，共 40 个座位。无论在哪个位置都可以看到大堂中央烧着热水的陶锅在冒着水汽，还能听到店员炒茶的声音，细细品味那扑鼻而来的茶香。这些都能愉悦五感。

Interiors are mainly made of woods. The ro, in the center of the store can be seen from most of the seats at the store, which comes to approximately 40 seats. You can enjoy with your 5 senses, such as the steam from kama and sounds and aroma of nihoncha being roasted.

搭配茶的水果与香草都是当季的，下单后店员会将茶叶以及将要使用的水果与香草放在精巧的果篮里带到客人面前，介绍制作的细节。

Fruits and herbs paired with nihoncha are selected in each season. A fruits and herbs basket is brought to your table together with tea leaves paired for them, and the staff explain the details.

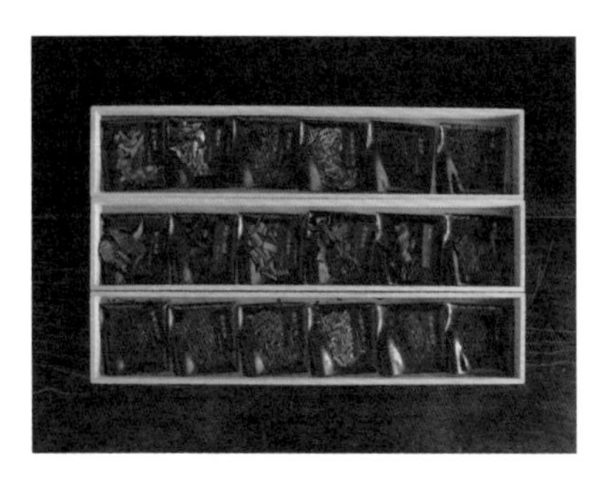

这是菜单上可点的茶叶的样本。有的茶叶可以直接冲泡，有的茶叶适合搭配水果与香草，店家会很乐意为您一一讲解。

Sample chaba used at HIGASHIYA. Some chaba can be enjoyed alone, and others are paired with fruits and herbs. Ask the knowledgeable store staff if you have any questions.

乍看之下以为是国外小报，其实是编排干练的菜单，菜单是日英双语的。除此之外，店内还有其他关于茶的书籍。

Menu designs remind you of overseas' tabloid papers. The menu is also written in English. There is also a book with detailed explanations on nihoncha available at the store.

不多加注意的话还会以为这件挂饰是个家徽，其实这是东屋茶寮的店标。还请您推开玻璃门，走进这个别具一格的日本茶空间吧。

The design of interior items reminds customers of kamon, symbolic Japanese family crest. Behind the glass door, you will find a new, Japanese tearoom space, one that you have never experienced before.

位于茶寮入口的收银台，整面墙都是原木色的隔层，犹如欧洲的杂货店一般，然而架子内排列错落有致的精美点心盒内都是和果子。

Beautifully packaged sweets are impeccably lined up on the shelf made of plain wood at the retail space beside sabou. It reminds you of a décor store in Europe, but at HIGASHIYA, the contents of the store are all wagashi.

手工最中（一种和果子）是此店的招牌，推荐您先将馅挤到细长的外皮内，将带着馅儿的两片外皮合在一起享用。共有黑芝麻馅、丹波大纳言红豆馅、芋头馅等 6 种口味，包装各不相同。1 盒内有两条，每盒 700 日元。

Monaka is boxed in a thin, candy bar like package. When you eat it, you will put an (red bean paste) in the thin, rectangular shaped Monaka skin. There are 6 kinds of Monaka available including Kurogoma An (black sesame), Tanba Tsubu An (red beans from Tanba region known for high-quality beans), Murasaki Imo An (purple sweet potato). Each kind is packaged in a unique colored box. 2 Monaka in 1 Box (700yen).

羊羹（一种和果子）有大小两个尺寸，分别有红豆馅、浓茶、椰子、焦糖 4 种口味。店家用印着店标的卷纸将羊羹精心包装起来，让您在打开包装的那一刻起就充满期待。大包装 1900 日元，小包装 1200 日元。

Large Yokan (1900 yen) & Small Yokan (1200 yen). Yokan comes in 2 sizes and 4 flavors, Dainagon Beans, Koi Cha (a dark green tea), Coconuts or Kogemitsu (a Japanese style caramel). Yokan is wrapped with a paper with HIGASHIYA's logo embossed on it and tied with a string, making it fun for the person opening the wrapping.

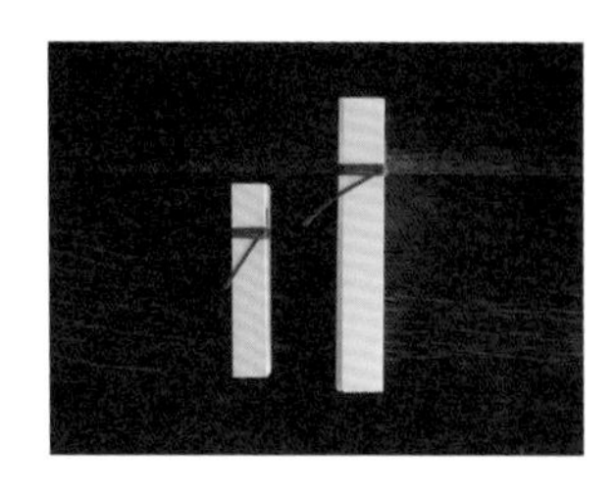

糖果纸包装起来的米花糖有荞麦、生姜、大德寺纳豆 3 种口味。复古风格的铁罐包装，1 盒 1400 日元；适合作为礼物的桐木箱包装，1 盒 1700 日元。

Okoshi in a tin (1400 yen) is individually wrapped like candy. They come in 3 flavors, Soba no Mi (barley seeds), Shoga (ginger), and Daitokuji Natto in a retro design tin. For a gift, Okoshi in Kiri Box (1700 yen) may be more appropriate.

有着东屋茶寮店标形状的落雁（一种和果子）也是送礼佳品，一共有麦粉、生姜以及竹炭这 3 种少有的新奇口味。方形的小盒包装，1 盒 1300 日元。

Rakugan in a Box (1300 yen). Rakugan in the shape of HIGASHIYA logo comes in 3 flavors, Roasted Oats, Ginger and Charcoaled Bamboo, all of which are rare and innovative Rakugan flavors. Rakugan comes in a cube-shaped box.

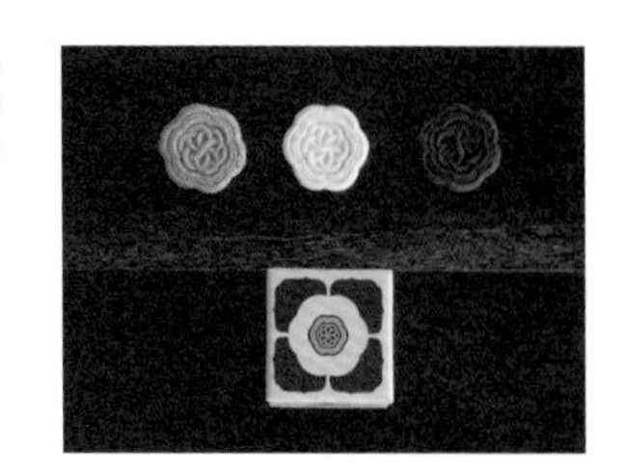

❺ 山本山日本桥本店茶室（日本桥）

山本山日本桥本店创立于江户时代元禄三年（1690 年），天明七年（1787 年）搬迁至日本桥，从那时起就是日本桥的一道风景，小店至今也在延续江户时代的风格。您可以悠闲地在展示各种茶器的店内挑选商品，也可以在连着店铺的休息区以及茶室享用日本茶与和果子，和果子来自日本桥的各个老店，比如“银装”的长崎蛋糕、“兔屋”的铜锣烧等。在茶室还可以品尝日本茶与和果子或者煎饼的套餐，这些茶点是来自日本桥老店“长门”的特供。除此之外，还有需要预约的茶泡饭或者海苔泡饭可供选择。若您在日本桥逛累了，还请步入这充满江户情怀的老店歇歇脚。

㊋时 10:00—18:00（6 月—7 月、11 月—12 月 营业至 19:00）㊋电 03-3281-0010 ㊋休 1 月 1 日 ㊋址 中央区日本桥 2-10-2 ㊋交 地铁银座线、东西线日本桥站 C4 出口步行 3 分钟 ㊋网 www.yamamotoyama.co.jp

❺ Yamamotoyama Nihonbashi Honten (Nihonbashi)

Yamamotoyama has been operating its store in Nihonbashi since it opened in 1787 (established originally in 1690 during the Edo era). Even today, it cherishes the concept of Edo Jocho, a nostalgic feeling for Edo. At the store, you can choose your items at your own pace. You can enjoy nihoncha and sweets at Okutsurogi Dokoro and Kissa Shitsu next to the store. Enjoy a set menu with local wagashi here, such as Castella from Ginso or Dorayaki from Usagiya. At Kissa Shitsu in the back, order a set menu of nihoncha and osenbei (rice crackers). Osenbei is from Nagato. This is a perfect place to take a break from strolling around Nihonbashi. Also, enjoy ochazuke, a simple Japanese dish in which hot tea is poured over cooked white rice topped with a few simple ingredients, with specially selected Nihoncha and Nori (Ochazuke by Reservation Only).

(H) Weekday:10:00-18:00 (June-July and November-December:10:00-19:00) (T) 03-3281-0010 (C) New Year's Day (Ad) 2-10-2 Nihonbashi, Chuo-ku (Ac) 3-minute walk from Nihonbashi Station Exit C4 (Tokyo Metro Ginza Line or Tozai Line) (U) www.yamamotoyama.co.jp

❻ 月光（莺谷）

月光是一家位于上野地区、提供日式年糕与日本茶的茶馆，店内洋溢着昭和时代的复古风情。店家注重制作日式年糕的糯米及其鲜度，并采用现在非常少见的手工制法，做出来的年糕口感丝滑而且容易下咽。在茶的泡制上，店内特意选用了爱知县生产的常滑烧茶壶，用来冲泡静冈牧之原生产的深蒸茶，由于深蒸茶尤为讲究第一泡的水温和闷蒸的时间，通常都是店主堀口亲自为客人泡制。第二泡之后客人可以自由地使用店内的热水，热水特意放在铁壶内，用这种含铁的热水泡出的茶会别有一番滋味（泡煎茶与芽茶的时候要将热水稍作冷却再泡）。月光这家店铺人气高的秘密就在于，客人在这里既可以吃到美味新鲜的年糕，又能续茶水。在这里还可以体验到纯正的京都抹茶。店内有英文说明书和菜单，在外国游客间有一定名气。

㊐ 12:00—19:30（星期六、星期日、节假日营业至 19:00，打烊前半小时不再接单）㊁ 03-5849-4569
㊡星期三 ㊟台东区根岸 3-7-18 El Alcazar 莺谷 1 楼 ㊋ JR 莺谷站北口步行 3 分钟
㊋minowa-gekko.com

❻ Gekko (Uguisudani)

Gekko is located near Ueno. The café, a specialty store of nihoncha and mochi (rice cakes), offers a nostalgic retro atmosphere, reminding customers of the Showa era. Gekko selects mochigome, a japonica glutinous rice, carefully maintains freshness, and always makes mochi hand-beaten, a rare technique used nowadays even in Japan. Gekko's mochi is smooth and soft. Nihoncha is Fukamushicha from Makinohara, Shizuoka. Kyusu is Tokoname Yaki, from Aichi, known for maximizing the nihoncha taste. The brewing method: the first cup is brewed by its owner, Mr. Horiguchi, because humidity and the length of time to steep must be carefully considered. From the second pot, visitors can refill their kyusu by themselves. When you refill, hot water from the iron pot contains iron, so you feel change in the taste of nihoncha. Enjoy nihoncha and mochi at your own pace. English versions of brochures and the menu are available, so Gekko is also popular among visitors from overseas.

Ⓗ Weekday:12:00-19:30 (Saturday, Sunday and Holidays:12:30-19:00) Ⓣ 03-5849-4569 Ⓒ Wednesday (Ad) El Alcazar Uguisudani 1st Floor, 3-7-18 Negishi, Taito-ku (Ac) 3-minute walk from Uguisudani Station North Exit (JR Line) Ⓤ minowa-gekko.com

❼ 寿月堂银座歌舞伎座店（银座）

寿月堂银座歌舞伎座店坐落在 2013 年开张的歌舞伎座大楼的 5 楼，拥有一处约 450 平方米的天台庭院，客人可以在庭院内享受茶品、和果子以及轻食。建筑师隈研吾在办公楼上打造了这样一处与世隔绝的茶空间，禅意十足地使用了约 3000 根竹子覆盖在天顶，而桌椅一律选择黑色的干练设计，制造出传统与现代风格的碰撞。由于茶室位于歌舞伎座，店内最受那些喜爱日本文化的观光客欢迎的是需要预约的“日本茶体验课”（3200 日元，英文服务则为 4000 日元）。体验时间为 1 个小时，您可以品尝到玉露、抹茶、焙茶等不同种类的茶品及搭配的茶点心，体验项目中还包括试吃茶叶、香炉品香等内容。伴手礼可以选择茶包上绘制了歌舞伎插画的茶套装，及其他一些歌舞伎座限定的商品。

㊐ 10:00—19:00（最后点单时间 18:30）㊂ 03-6278-7626 ㊡ 全年无休 ㊟ 中央区银座 4-12-15 歌舞伎座大楼 5 楼 ㊋ 地铁日比谷线东银座站直达 ㊓ www.maruyamanori.com

❼ Jugetsudo Ginza Kabukiza (Ginza)

Jugetsudo Ginza Kabukiza is located at a spacious roof garden on the top floor of the newly opened Kabukiza in 2013. Enjoy nihoncha, sweets and light foods in a luxurious space with garden view. Designed by Kengo Kuma, the store is covered with 3,000 bamboos creating the world of Zen. It offers a contrast between mainly a black-colored theme inside the store and the bamboos. Many tourists visit this extraordinary space probably due to the fact that it is located in Kabukiza. "Tea Experience Menu" is particularly popular (3,200 yen including tax. 4,000 yen including tax for the course in English. Reservation Only). This 1-hour course includes Gyokuro, Matcha, Sencha and Hojicha paired with sweets. Make sure to allocate enough time to eat chaba, and experience Hojicha roasting at a chakou ro, a nihocha burner. Various limited edition items, exclusively sold at Kabukiza Store, are available, such as nihoncha in tea bags decorated with Kabuki illustrations.

Ⓗ 10:00-19:00 (L.O.18:30) Ⓣ 03-6278-7626 Ⓒ None (Ad) Kabukiza Tower 5th Floor, 4-12-15 Ginza, Chuo-ku (Ac) Direct Connection to Higashi Ginza Station (Tokyo Metro Hibiya Line) Ⓤ www.jugetsudo.fr

❽ 茶茶工房（早稻田）

茶茶工房靠近高田马场站，距新宿附近的热闹的街道稍有距离，白色的外墙前摆满了绿色植物，手写的木制招牌显得特别亲切。店内家具以黑色及木制为基调，墙上的挂钟钟摆仿佛让时间都慢了下来。小店于 2003 年开张，从开店起店家就坚持只提供无农药及有机栽培的茶，茶品精选自全国各地，菜单标注了各类茶的香气、甜味以及涩味的程度。吧台上摆放着铁壶与舀水的竹制长柄勺子，店家基本都会用铁壶烧制的热水为客人泡上第一道茶，并会教客人一些泡茶的基本方法，之后便让客人自己慢慢享用。这里是一个可以让人放松久坐的空间，也是一个当地人都不时会去小坐一下的地方。店铺营业至晚上十点，深夜还会提供饭团、盖饭、轻食以及酒精饮料。

(时) 12:00—22:00 (电) 03-3203-2033 (休) 星期日、节假日 (址) 新宿区西早稻田 2-21-19 (交) JR、西武新宿线、地铁东西线高田马场站步行 7 分钟 (网) chachakoubou.com

❽ CHA CHA KOUBOU (Waseda)

Located near Takadanobaba Station, CHA CHA KOUBOU welcomes visitors with a warm, hand-written wooden sign. Once inside, the store has a quiet atmosphere mainly designed with black and wooden materials, with the tick tocking sound of a wall clock. Since it opened in 2003, the café offers organic, chemical-free nihoncha. CHA CHA KOUBOU's menu indicates each tea's level of aroma, sweetness and shibumi. At the counter, you will find an iron pot and hishaku (a ladle). Nihoncha is carefully brewed with hot water from the iron pot. Your order is accompanied by a pot with hot water for a second brew, and don't worry, because the staff will guide you on how to brew a good second cup. You may lose track of your time, because of the comfort it provides. At late night hours, local residents visit the café for light foods, such as onigiri (rice balls), and over-rice dishes together with alcoholic beverages.

Ⓗ12:00-22:00 Ⓣ03-3203-2033 Ⓒ Sunday and Holidays (Ad)2-21-19 Nishi-Waseda, Shinjuku-ku
(Ac) 7-minute walk from Takadanobaba Station (JR Seibu Shinjuku Line or Tokyo Metro Tozai Line)
Ⓤ chachakoubou.com

店内提供的茶点为水羊羹（500 日元），照片上使用的茶叶——“天丰”产自鹿儿岛，是一种一根茶树枝干上只能摘下 5 ~ 6 片新叶的稀有茶品，只在 5 ~ 9 月之间提供。

Mizu-Yokan (500 yen). Nihoncha in the photo is Tenhou, Mushicha from Kagoshima. Tenhou is made with new tea leaves. Only 5 or 6 tea leaves grow on one tea tree. Tenhou is served for a limited time between May and September.

墙上的木架上摆放着贴了商店标签的铁皮盒装的各类茶叶，店内装饰的小物以和风为主，给人一种在日式老房子里的安心感。菜单上的所有茶品都可以买到相应的伴手礼。

The space offers a feeling of being at a kominka, an old Japanese-style house, with small Japanese style décor items such as chazutsu, a tea caddy with the store name label on it, displayed on wooden shelves. You can purchase chaba on the menu as well.

其他推荐
BEST CAFES and MORE

虎屋东京店（丸之内）

店内保留了东京站初期的红色墙砖。

(时) 10:00—21:00（星期日、节假日营业至 20:00）(电) 03-5220-2345 (休) 全年无休 (址) 千代田区丸之内 1-9-1 东京站酒店丸之内南广场 2 楼 (交) JR 东京站丸之内南口直达，乘坐酒店内电梯上至 2 楼 (网) www.toraya-group.co.jp

TORAYA TOKYO (Marunouchi)

Red bricks of old Tokyo Station are well-incorporated in the interiors of the store.

(H) 10:00-21:00 (Sunday and Holidays:10:00-20:00) (T) 03-5220-2345 (C) None (Ad) Tokyo Station Hotel, South Dome 2nd floor, 1-9-1 Marunouchi, Chiyoda-ku (Ac) Tokyo Station, Marunouchi South Exit (JR Tokyo Station). Take the elevator up to the 2nd floor of the Hotel. (U) www.toraya-group.co.jp

---

林屋新兵卫（银座）

京都老店“京林屋”在东京新开设的茶咖啡酒吧。

(时) 11:30—21:00 (电) 03-6280-6767 (休) 不定期 (址) 中央区银座 7-4-5 银座 745 大楼 1 层 (交) 地铁银座站 B5 出口步行 5 分钟 (网) www.hayashiya-shinbei.com

HAYASHIYA SHINBEI (Ginza)

Kyo Hayashiya's new expansion, nihoncha café & bar.

(H) 11:30-21:00 (T) 03-6280-6767 (C) Irregular Closure Days (Ad) Ginza 745 Building 1st Floor, 7-4-5 Ginza, Chuo-ku (Ac) 5-minute walk from Tokyo Metro Ginza Station Exit B5 (U) www.hayashiya-shinbei.com

---

茶 CAFE 竹若（银座）

在玻璃门的屋内享用各类茶品。

(时) 11:00—22:30 (电) 03-6264-7585 (休) 全年无休 (址) 中央区银座 4-10-5 东急 Stay 银座店 1 层 (交) 地铁日比谷线东银座站步行 1 分钟 (网) www.takewaka.co.jp/chacafe

Cha Café Takewaka (Ginza)

Enjoy various nihoncha at the spacious glass-fronted café.

(H) 11:00-22:30 (T) 03-6264-7585 (C) None (Ad) Tokyu Stay Ginza 1st Floor, 4-10-5 Ginza, Chuo-ku (Ac) 1-minute walk from Higashi-Ginza Station (Tokyo Metro Hibiya Line) (U) www.takewaka.co.jp/chacafe

---

茶之叶（银座）

一边欣赏绿植，一边感受茶与休闲时刻的美好。

(时) 10:00—20:00 (电) 03-3567-2635 (休) 随松屋银座的营业时间 (址) 中央区银座 3-6-1 松屋银座 地下 1 层 (交) 地铁银座站 A12 出口直达 (网) www.chanoha.info

CHANOHA (Ginza)

Cherish greenery and enjoy your break with nihoncha.

(H) 10:00-20:00 (T) 03-3567-2635 (C) Same hours as Matsuya Ginza's operating hours (Ad) Matsuya Ginza B1 Floor, 3-6-1 Ginza, Chuo-ku (Ac) Tokyo Metro Ginza Station A12 (Tokyo Metro) (U) www.chanoha.info/english

第二章

品茶：
了解日本茶的分类与特色

ENJOY NIHONCHA:
EXPLORE ITS VARIETIES

若想了解每家店的特色，请选择店家独一无二的菜单

位于表参道的茶茶之间茶店是由茶艺师和多田喜于十二年前开设的，也许是职业的影响，和多田喜对于茶有着自己的坚持。开业以来的十二年间，茶茶之间一直专注于使用“出自同一茶园的同一品种”的茶叶。和多田喜认为即使是同一种茶叶，每浸泡一次后呈现出的香气与味道也不相同，甚至茶器的变化也会对其味道有所影响。同样，他在对比了许多不同茶园及品种的茶叶后发现，即使泡法类似，每种茶所呈现出的色香味也大不相同。和多田喜说：“在他刚开店的时候，日本人还没有要到茶店付钱喝茶的这种意识，现在就大不相同了。”东京的街道上出现了各式各样的日本茶店，每家店都会用心地打造自己的招牌茶饮以及提供独特的喝法。还请您亲自去探寻最适合自己的日本茶吧。

Experience a beautiful cup of nihoncha only available at each store and explore its unique menu.

It has been 12 years since Yoshi Watada, a Nihoncha Sommelier, opened his café, Omotesando chachanoma (Omotesando at page 59). His policy is to serve single-origin nihoncha, a single kind of chaba grown on a single farm. Yoshi searched for a beautiful cup of nihoncha to reach this policy. “The taste of the same nihoncha will change between the first cup, second cup, third cup and so on.” “Compare each farm, each kind of chaba. I use the same brewing techniques, but different kinds of chaba. You will be surprised by the unique characteristic that each nihoncha brings to you.” Yoshi says. “12 years ago when I opened chachanoma, people felt awkward to be paid for nihoncha. It has changed now. The way both customers and café owners think, has changed.” Yoshi explains. There are so many nihoncha cafes in Tokyo now, serving not only a wide variety of chaba, but also a wide variety of how you can drink your nihoncha, all arranged in a unique way. Explore and find the café that fits your own preference.

ARTRIP ADVISER
艺术之旅顾问

和多田喜
Yoshi Watada

日本茶茶艺师，出生于东京。表参道茶茶之间茶店的店主，著有《日本茶茶艺师和多田喜教你从今天开始学会品茶》一书。

Nihoncha Sommelier. Born in Tokyo. Owner of Omotesando chachanoma. His books include *Nihoncha Sommelier, Yoshi Watada no Kyokara Ocha wo Oishiku Tanoshimu Hon：A Book for Enjoying Good Nihoncha from Today by Nihoncha Sommelier, Yoshi Watada*.

❾ 櫻井焙茶研究所（表参道）

樱井焙茶研究所位于表参道 Spiral 商场 5 楼。下了电梯往右走就可以看到一道玻璃门，阳光穿过通透的玻璃射进店里，明明是茶店，却有着研究所般的沉静氛围。“研究所所长”樱井会使用专门的机器烘焙茶叶，还会精选来自全国各地的茶叶亲自配比、混合、烘焙，制作配方茶，专注于“研究”各种日本茶的新喝法。在店内吧台可以点到店内特制的普通蒸煎茶、焙茶等由 10 种不同制作方法制成的茶品，每杯 400 日元。靠里的茶房有黑色 U 形桌，一共有 8 个座位。菜单上有煎茶、焙茶等日本茶与茶点的套餐，还有店家独创的“茶酒”。即使是同一种茶，烘焙的程度不同、混合的茶叶种类的不同，呈现出来的风味也是不同的。这就是樱井焙茶研究所最具魅力的“研究成果”。

(时) 11:00—23:00（最后点单时间 22:30）(电) 03-6451-1539 (休) 依照 Spiral 商场的休息时间 (址) 港区南青山 5-6-23 Spiral 商场 5 楼 (交) 地铁表参道站 B1 出口直达 (网) www.sakurai-tea.jp

❾ SAKURAI JAPANESE TEA EXPERIENCE (Omotesando)

SAKURAI JAPANESE TEA EXPERIENCE is located on the 5th floor of Spiral in Omotesando. The store probably reminds you of a Research Center. Inside, the director of the SAKURAI JAPANESE TEA EXPERIENCE, conducts his research, utilizing the specially designed tea roaster to roast tea, or blending chaba, which are procured from all around the world. At the counter by the entrance, 10 varieties of teas, such as Futsumushi-Sencha and Hojicha, can be enjoyed in a cup for 400 yen each. In the back, there is a tea salon with a black, U-shaped counter with 8 seats. Its menu includes nihoncha and wagashi set, a course with varieties of teas paired with ochauke (small sweets), made to be enjoyed with nihoncha or original chashu (nihoncha sake). By the time you leave this Research Center, you will recognize the charm of nihoncha. Discover new ways to enjoy nihoncha by noticing changes in chaba characteristics created by the degree of roasting or the items paired or blended with it. It offers indefinite possibilities.

Ⓗ 11:00-23:00 (L.O.22:30) Ⓣ 03-6451-1539 Ⓒ Same hours as Spiral's operation hours (Ad) Spiral 5th Floor, 5-6-23 Minamiaoyama, Minato-ku (Ac) In front of Omotesando Station, Exit B1 (Tokyo Metro) Ⓤ www.sakurai-tea.jp

店内的装潢主要选择原木与黄铜一类的材料，这样的材质让人仿佛可以看到静物随着时间流逝而产生的岁月痕迹。樱井所长与店内的服务员都身着白大褂，仿佛就是“研究所”里的工作人员。

Copper and Wood are mainly used for interiors, aiming to enjoy how they change colors as the years go by. Director Sakurai and other staff wear white robes, which also suits the concept of Research Center.

烘焙室是一个被玻璃包围起的空间。心灵手巧的店主将咖啡烘焙机改造成茶叶烘焙机，无论是怎样的烘焙方式——浅度烘焙、中度烘焙，还是深度烘焙，樱井所长都在致力于“研发”出茶叶最好的口感。

Baisenshitu, literally means roastery in Japanese, is covered with glass walls. The research of the best roasting method to extract tasty nihoncha from each chaba, is conducted with an original roasting machine, which is specially designed for the store. Methods include asairi (light roast), chuiri (medium roast) and fukairi (deep roast).

这幅书法作品由书法家铃木猛利书写。一个“茶”字，被端正地摆放在展示茶叶及茶器的木架上。

The Japanese character, cha, is written by Mori Suzuki, a calligrapher. It sits on the wooded shelf where chaba and chaki are displayed for sale.

茶房的主桌上有一个以“蹲踞”为设计原型的洗手池。店内狭窄的过道、烧热水的陶锅以及静谧的环境，营造出身处日式茶室的感觉。

On the counter at the tea salon in the back of the store, there is a washbasin, which reminds visitors of tsukubai, a stone washbasin typically placed near the entrance of a chashitsu, a traditional Japanese tearoom. A thin, narrow hallway, a chagama, a teapot and its tightness in the air let the visitors feel the atmosphere of chashitsu at sabou.

“研究所”室内外到处都是这种让人联想到东方药罐的古董壶，营造出“研究所”以及“实验室”的气氛。

An antique pot from the East. It makes us think of a medicine pot used in the East during the old days. Items reminding visitors of the research center or the laboratory are placed around both inside and outside of the rooms.

樱井焙茶研究所的招牌：日本茶深度体验套餐

A COURSE MENU FOR EXPERIENCING THE DEPTH OF NIHONCHA

① 第一道玉露茶

为了让茶叶的香气更加均匀地散开，店家特意使用扁平的小锅。第一道茶的味道较重，搭配简单调理过的红豆最为合适。一口一粒的红豆搭配口感厚重的玉露茶，两种味道可谓相得益彰。

① GYOKURO, 1ST CUP

To open the chaba and extract umami, houhin (a flat pot), is used.
1st cup has a deep umami taste, so served in a small portion.
Plainly cooked, azuki (red beans), are served with the cup. Slowly bite each bean. Azuki goes well with rich gyokuro.

② 第二道玉露茶

用比第一道温度更高的热水泡茶，因此茶叶的涩味和苦味会被更明显地释放出来。搭配的小点心是沾了大豆粉的“州浜”，味道偏甜，更适合搭配第二道玉露茶。

② GYOKURO, 2ND CUP

A 2nd cup is brewed with hot water at a higher temperature than the 1st cup.
It has more shibumi taste, so tsukeawase (sweets for tea), taste sweeter with teas.
Tsukeawase is a sweet called Suhama with Uguisu Kinako.

③ 第三道玉露茶

将第三道玉露茶搭配当季香草制成冷泡茶。若在初夏的时候会将清爽的酸橘与紫苏叶放进饮品里，口感清爽。这一道茶搭配的是将冲泡后的茶叶用咸柚子醋腌制而成的爽口配菜。去享受这浓郁的美味吧。

③ GYOKURO, 3RD CUP

A 3rd cup is a cold tea blended with seasonal herbs.
For summer, gyokuro is blended with sudachi and purple shiso leaves, offering a tea truly refreshing.
Tsukeawase is chaba, served with shio-ponzu - citrus soy sauce. Enjoy concentrated umami of chaba.

④ 焙茶

接下来端上的是由客人自己选择的茶叶现做的焙茶，搭配的小菜是醋腌小碟。与玉露的泡法不同，焙茶推荐使用茶壶冲泡。

④ HOJICHA

Hojicha, roasted to match your taste on the spot, is made with selected chaba, utilizing horoku (a tea roaster). Tsukeawase for hojicha are ko-no-mono, Japanese pickles. Simple pickles go well with hojicha with a roasted flavor. Tea is brewed with a kyusu.

⑤ 薄茶与和果子

套餐的最后一道是薄茶与和果子（照片中为板栗羊羹）。这个深层体验套餐为 4500 日元，可以在里屋的茶房里边欣赏服务员泡茶，边细细品味。

⑤ OUSU AND WAGASHI

The course (4,500 yen) ends with ousu and wagashi.
(In the Photo: Kuri Yokan)
Visitors can enjoy this course at the back of the tea salon while viewing the matcha whisking techniques.

和果子的种类

搭配薄茶的和果子是小布施产的板栗羊羹、板栗团子以及“东屋”系列店特供的当季生果子。

WAGASHI SELECTIONS

You can choose which wagashi to enjoy with ousu, such as Kuri-Yokan, Kuri-Kanoko, or fresh seasonal sweets of HIGASHIYA, which is a sister store.

⑩ 东京茶寮（三轩茶屋）

东京茶寮于 2017 年开张，是日本国内第一家将手冲咖啡与日本茶相结合的新概念“茶吧”。在这里您可以享受到不同产地、品种以及不同制法的煎茶。店主以手冲咖啡的理念为基础，使用特制的手冲器具来冲泡茶叶。小店有 9 个位置，客人无论坐在吧台前的哪个位置都可以看到店员泡茶的过程，这种可以跟周围人闲聊类似咖啡店的轻松氛围，其实是为了致敬茶道的茶会中，亭主打好一碗浓茶后，传递给所有来客，大家一人一口共品一碗茶的做法。这个略显狭窄的空间里，陌生人之间的疏离感会因为共享这样的轻松氛围而减缓，给人以放松的感觉。这是一家以崭新的泡茶方法引领品茶新潮流的茶店。

㊐ 13:00—20:00（星期六、星期日、节假日 11:00—20:00）㊔ 无 ㊡ 星期一（节假日顺延）㊟ 世田谷区上马 1-34-15 ㊋ 东急田园都市线三轩茶屋站步行 7 分钟 ㊍ www.tokyosaryo.jp

⑩ TOKYO SARYO (Sangenjaya)

Tokyo Saryo opened in 2017. The café offers carefully selected sencha, steamed or roasted masterfully at just the right level. Its Tea Flight invites you to enjoy the uniqueness of each nihoncha. It serves the world's first hand-dripped, single-origin sencha, using the sencha drip tools it invented. The atmosphere is similar to that of a café serves special coffees, as you get to enjoy conversations with the barista and other customers. In fact, this is also connected to a core philosophy of sado, teachings associated with traditional tea ceremony, in which all people at the ceremony enjoy matcha in a single bowl together. The café offers a minimalistic space to enjoy the moment together. Catch the new "Third Wave" of nihoncha here.

(H) 13:00-20:00 (Saturday, Sunday and Holidays:11:00-20:00) (T) None (C) Monday (If Monday is a Holiday then closed on Tuesday) (Ad) 1-34-15 Kamiuma, Setagaya-ku (Ac) 7-minute walk from Sangenjaya Station (Tokyu Denentoshi Line) (U) www.tokyosaryo.jp

客人可以选择两种茶叶，店家会泡上两道后一起端上来，客人可以选择其中的一种茶叶，要求店员混上炒好的玄米制成玄米茶后品尝。此图中左边的是产自鹿儿岛的“Harumoegi”，右边的是静冈县产的“香骏”。靠图片上部的两杯是第一道，用了像红酒杯一样的茶杯盛放，是为了让茶香更容易散发出来；靠前的两杯则是偏高温的第二道，所以用的是有把手的茶杯，防止客人烫手。

Enjoy 2 cups each with 2 varieties of chaba. For one of the chaba, try the third cup as genmaicha.
Left, Harumoegi. Right, Koushun. The cup in the front is a wine-glass shaped 1st cup. The 2nd cup has a handle.

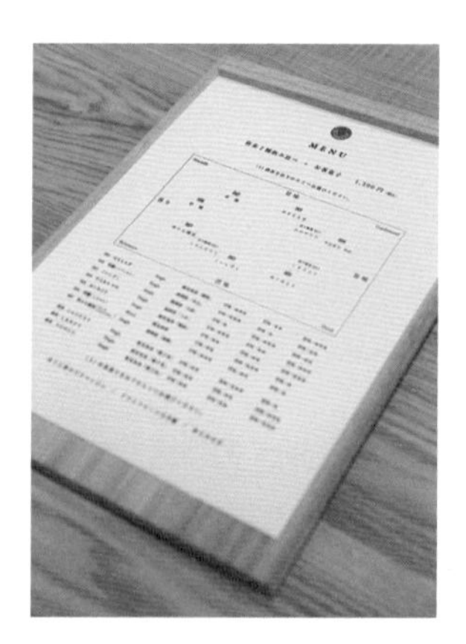

图中为茶饮的菜单，上面标明了茶叶的产地、甜度、香度、涩味、回甘程度等作为参考，套餐组合中客人可选择两种茶，搭配的茶点有果干、焙茶和牛奶制成的奶茶冻以及萩饼，一套组合 1300 日元。

Nihoncha menu. Select 2 varieties of chaba for Tea Flight, based on its origin, degrees of amami, aroma, shibumi and umami. Select one of the sweets, Dried Fruits, Hojicha Blancmange or Kaoru Ohagi for a set. (1,300 yen, tax included)

除去一小块被店家暖帘遮挡住的储物空间外，店内的设计基本是以让客人能够全情体验日本茶为理念。

Except a small backyard space behind the noren with logo designs printed on, the interiors of the store are designed with the concepts to show and have visitors experience.

店内提供的茶叶都可以购买，为了保持茶叶的新鲜度，每袋包装都填充了氮气。包装上还印着茶叶产地、茶园、海拔、蒸法、烘焙火候以及甜味、苦味、回甘的程度等信息。

Chaba sold at the store is in a nitrogen-filled package, which keeps it fresh. Each package indicates origin, tea farm, altitude, mushi (steaming process), and hire (roasting process information). It also indicates the degrees of amami, nigami and umami.

店内销售的如茶叶罐等保存茶叶的器具，在设计上都考虑到了客人的使用感，希望客人可以享受泡茶的每一步过程。今后店家还计划推出茶器、筛选器具等物件作为商品销售。

All the items used at the store, such as chazutsu, are designed while being conscious about being viewed by visitors in a process of tea brewing and serving. Tokyo Saryo plans to make selected items, including chaki and hand-drip sets available for sale.

东京茶寮的特色：两茶三喝

TASTE 2 VARIETIES OF CHABA IN 3 WAYS

① 称量茶叶

一人份需要 4 克茶叶，店家不使用电子秤，选用传统的天平称重。用沙漏来计算浸泡时间。

① MEASURE CHABA

4g of chaba is used for 1 person.
Tokyo Saryo does not use any digital device for its tea brewing process.
To measure chaba, a balance scale is used. To time the brewing process, a sandglass is used.

② 倒入茶叶

倒入茶叶之前先将器具摆好，将手冲咖啡会用上的漏壶以及滤网架在杯子上方，然后将称量好的茶叶倒在滤网上。

② PUT CHABA IN

Set a dripper and filter over a cup for serving.
Put chaba in the filter.

③ 注入热水

每一次注入的热水为 120 毫升，泡第一道时的水温为 70 摄氏度，第二道的水温为 80 摄氏度。第一道茶主要在于能让茶叶的香气与甘甜之味（氨基酸）被激发而出；第二道用较高温度的热水，意在将茶的苦涩味（咖啡因与儿茶酸）冲泡出来。

③ POUR OVER HOT WATER

120 mℓ of hot water at a temperature of 70 degree Celsius for the 1st cup.
80 degree Celsius for the 2nd cup. The 1st cup extracts aroma and amami (amino acid). The 2nd cup extracts shibumi and nigami (caffeine and catechin) by increasing the water temperature.

④ 萃取

倒入热水之后，第一道茶泡上 90 秒，第二道茶则只需泡 15 秒。因为漏壶与滤纸的尺寸都经过精密的测算，在等待期间，热水会完整地包裹着茶叶，浸泡并激发出茶叶的味道，不会滴漏进杯子里。

④ EXTRACTION

After hot water is poured in, wait for percolation for approximately 90 seconds for the 1st cup and 15 seconds for the 2nd cup. During this time, hot water barely drips into the cups, because the sizes of the dripper and filter are calculated and designed accordingly.

⑤ 冲泡

将漏壶稍作倾斜，茶水沿着杯壁流入杯中，光是观看茶水流入杯中的过程也不失为一种享受。

⑤ BREW NIHONCHA

The dripper needs to be tilted in order to release all nihoncha into the cups for serving.
The visual image of the green nihoncha dripping through the filter will captivate you.

⑥ 泡制玄米茶

在泡了第二道的茶叶上放入少许炒好的玄米来制作第三道玄米茶。注入 80 摄氏度的热水静置 15 秒，滤掉茶叶后便可以享用。

⑥ BREW GENMAICHA

After a 2nd cup is brewed, add genmai rice and extract genmaicha as a 3rd cup.
Hot water at a temperature of 80 degree Celsius is poured over. Wait for approximately 15 seconds for the 3rd cup to completely drip through the filter into the cup.

⓫ 茶茶之间（表参道）

“单一品种”一词原指单一产地的咖啡豆，而位于表参道的茶茶之间茶店将这个词引入茶道中，并延伸出“出自同一茶园的同一品种”的意义。茶茶之间于 2005 年开张，店主和多田喜亲自访问众多茶园产地，从各地的农家或者茶铺收集了近 30 种日本茶，尝试各种方法让每种茶的味道特点都能被最大化地呈现出来。客人只要坐在店内便可以看到茶艺师熟练地冲泡每一道茶品，第一道选用冰水萃取出茶叶的精华之味，第二道加入热水引出茶叶的涩味，第三道用冷水激发香气，第四道则用热水再次唤醒茶叶的清香之味，第五道茶水则是起到为客人润喉的作用。边欣赏茶艺师精湛的技艺，边细细品味着手中的茶，客人眼前仿佛浮现出片片茶田的自然之景。希望你可以亲自去体验一番“出自同一茶园的同一品种”的好茶。

(时) 11:00—19:00（最后点单时间 18:00）(电) 03-5468-8846 (休) 星期一 (址) 涩谷区神宫前 5-13-14
(交) 地铁表参道站步行 5 分钟 (网) www.chachanoma.com

⑪ Omotesando chachanoma
(Omotesando)

The term single-origin is generally used for coffee grown within a single known geographic origin. At chachanoma in Omotesando, only single-origin nihoncha, a single variety of chaba, grown on a single farm has been served since 2005. Its menu offers over 30 varieties. Café owner, Yoshi, goes to the tea farms and only procures chaba from tea farmers or tea retailers that he trusts. He brews all nihoncha with the best method for the tea to maximize its unique character. The first cup is a cold brew with water and ice to enjoy the essence of the nihoncha. The second cup is brewed with hot water to enjoy shibumi. The third cup is brewed with cold water to enjoy the aroma again. The fourth cup is brewed with hot water again to enjoy again the unique aroma. The fifth cup is brewed for quenching your thirst. Customers can view the open counter from their seats, where the Nihoncha Sommelier will brew your tea with style and great care, cup by cup. Picture nihoncha tea farm in your mind as you drink your cup at your own pace.

Ⓗ 11:00-19:00 (L.O.18:00) Ⓣ 03-5468-8846 Ⓒ Monday (Ad) 5-13-14 Jingumae, Shibuya-ku
(Ac) 5-minute walk from Omotesando Station (Tokyo Metro) Ⓤ www.chachanoma.com

秋津岛套餐（1500日元）。在这份套餐中，客人可以品尝到茶叶的五次冲泡，每次冲泡使用的茶器都尽量呈现茶的风味，很值得玩味。不仅是喝茶，套餐还会附上一小撮茶叶，供客人品香或是直接食用。

Akitsushima (1,500 yen).
You can enjoy 5 cups of Akitsushima, each cup brewed utilize the method to maximize the character of the tea with carefully selected chaki. Chaba will be served with the tea brewed as well, so you can eat them or enjoy the aroma.

三味茶拼盘（800 日元）。三味茶套餐中的茶品会根据季节变更，此套餐中的茶叶不使用茶壶冲泡，而是直接在茶杯里注入热水。仅是如此操作，品质优良的茶叶便能散出沁人心脾的香气。这也是店铺内的人气套餐茶品。
※现在改成了用茶壶冲泡的“两味茶”套餐。

Tea Flight Set with 2 kinds of seasonal nihoncha (800yen). A kyusu will not be used for brewing as the hot water will be directly poured on to the chaba, in the chahai (tea cup). The high-quality nihoncha will show its character just by that. This set is unique to chachanoma and is very popular.

明亮的店内摆放着原木色基调的家具，墙上装饰的是屏风画师岛田由子特意为招牌“秋津岛”的产地茶园所绘制的作品《茶之乡》。

The shop interior is bright and comfortable with glass windows and wooden interiors.
The picture on the wall is created by Fusumaeshi Artist, Yoshiko Shimada, titled “Cha no Sato - Home of Cha,” which depicts the tea farm of Akitsushima, the signature nihoncha served at chachanoma.

“茶茶之间”，取自“能够品茶的茶间 ”之意。店内可购买茶叶以及茶壶，既适合自家使用，也是送礼佳品。

The name, chachanoma, originates from chanoma, a living room where you can enjoy nihoncha in Japanese. So, at chachanoma, visitors can purchase chaba or kyusu that are used at the store, and bring them back and enjoy nihoncha at home. They are perfect as a seasonal gift too.

窗边有一面混凝土墙，有时候会挂着收获新茶的照片，有时也会被用作小众艺术家的摄影作品展示墙。

A bare concrete wall is used as an exhibition space.
Photos of nihoncha harvesting or art pieces from an artist are exhibited here.

伴手礼的包装也是别出心裁。店家根据茶的香气、味道等特色给茶取了不同名字，再以名字为创意灵感设计了不同的包装。标签上还注明了茶的甜味、甘味、涩味以及香味的程度。

Each tea is named based on its unique characteristic of taste and fragrance. Also, the tea packaging is unique and designed by visualizing the name of each tea. The characteristics of the tea are indicated by amami (sweetness), umami, shibumi and zankou (remaining aroma).

⑫ Tsukimasa（下北泽）

在被年轻人包围的下北泽街道上，有一处当地居民热爱的老茶馆“Tsukimasa”，它于1978年开业。进入店里，首先看到的是茶叶的销售处，里屋才是喝茶的地方，Tsukimasa家的茶品十分丰富，其中也有许多自制的原创饮品。店招牌饮品当属“梅子生姜番茶”，生姜对治疗感冒有效果，因此店家特意将生姜搭配梅子制成了这款招牌茶饮。另一款值得推荐的是“黄油茶”，这款茶来自于店主相马先生在西藏旅行时得到的灵感，做法就是将黄油与食盐放入番茶中。除了特色茶品外，也提供普通的煎茶、玄米茶、焙茶、玉露以及抹茶等。店主会根据客人的需求和心情推荐茶饮，这种轻松的氛围大概也是这家老店维持多年的秘诀了吧。照片中的煎茶套餐包含一份糕点，可随意添加热水，550日元一位。其他的茶饮价格也十分亲民，店家会教您如何泡出一杯味道好的茶饮，这是一家会让人流连忘返的茶店。

㊍11:00—21:00 ㊇03-3410-5943 ㊡周三 ㊟世田谷区代泽5-28-16 ㊣小田急线、京王井之头线下北泽站南口步行5分钟 ㊙www.tukimasa-simokita.com

⓬ Tsukimasa (Shimokitazawa)

Tsukimasa is located in Shimokitazawa, and has been loved by its community since 1978. By the entrance, chaba and other items are sold at a retail space. A café, located in the back, is known for its unique, wide menu selections. Original tea menu includes, Ume Shoga Bancha, nihoncha with plum and ginger, inspired by shoga-yu, a grated ginger in hot water, commonly served when people have a cold. Butter Cha, tea with butter and salt, is inspired by the café owner, Mr. Aiba's experience in Tibet. You can also enjoy standard nihoncha (sencha), genmaicha, hojicha, gyokuro or matcha. Enjoy a cup of tea depending on their mood that day. Sencha Set (550 yen) is served with hot water for refill and ochauke sweets. Prices are reasonable and all teas are brewed with care. Visitors can also learn how to brew teas. For many, this store is ikitsuke, a store people regularly go to.

(H) 11:00-21:00 (T) 03-3410-5943 (C) Wednesday (Ad) 5-28-16 Daizawa, Setagaya-ku (Ac) 5-minute walk from Shimokitazawa Station, South Exit (Odakyu-Line and Keio-Inokashira-Line) (U) www.tukimasa-simokita.com

⑬ Artless 手冲茶和咖啡店（中目黑）

Artless 手冲茶和咖啡店开业于 2016 年，店主川上旬是公关公司 artless Inc. 的老板，个人有收藏茶具的喜好，他的收藏包括国宝级艺术家高桥敬典制作的陶锅、常滑烧茶壶、开化堂八木隆裕制作的茶叶罐以及如同艺术品般的金属滤茶网等等。2017 年，这家店与公关公司一同迁移到了中目黑高架桥下，店内颇为显眼的是那张长达 5 米的铁质长桌，还有摆放在木架子上的各类茶具。正如店名中的“手冲”二字，店内所用的茶与咖啡都是来自小规模农场，都是有机无农药栽培出的“单一产地”产品。日本茶有 4 个种类供选择：无农药有机栽培的三重焙茶、京都炒番茶、宇治的无农药有机栽培玄米茶以及京都玉露茎茶。当客人点上一壶茶，茶艺师会在客人面前细致地冲泡，还请在那茶香中无限地遐想每款茶背后的故事。

㊐11:00—20:00 ㊉03-6434-1345 ㊡不固定 ㊟上目黑 2-45-12 中目黑高架桥下 85 ㊋东急东横线、地铁日比谷线中目黑站步行 5 分钟 ㊕www.craft-teaandcoffee.com

⑬ artless craft tea & coffee (Nakameguro)

In 2016, Shun Kawakami, the managing director of the branding agency, artless Inc., opened a shop specializing in nihoncha and coffee, adjacent to his office in Harajuku. He opened the store as an extension of his hobby of collecting chaki. In 2017, artless Inc. moved to a new location in Nakameguro under the viaduct. artless Inc. also brought iconic interiors from the initial location when they moved, such as chaki and its symbolic counter, made with a single iron plate 5 meters in width. It selects nihoncha and coffee from small organic or non-chemical farms, where chaba and coffee beans are grown with care. All selected nihoncha and coffees are single-origin. 4 kinds of nihoncha are available: organic, non-chemical farming, hojicha from Mie; karibancha from Kyoto; hand-roasted, organic, non-chemical farming, genmaicha from Uji; and kuki-gyokuro from Kyoto. Barista will brew your tea with care. Imagine the story behind your nihoncha while enjoying your cup.

(H) 11:00-20:00 (T) 03-6434-1345 (C) Irregular Closure Days (Ad) Nakame Gallery Street J2, 2-45-12 Kami-Meguro (Nagameguro Koka-Shita 85) (Ac) 5-minute walk from Nakameguro Station (Tokyu-Toyoko Line and Tokyo Metro Hibiya Line) (U) www.craft-teaandcoffee.com

⑭ 幻幻庵（涩谷）

幻幻庵的店铺让人感觉十分新潮：全黑的外墙，全英文的手写招牌，店内播放的还是美国黑人的音乐。但它可是一家地道的日本茶店，经由日本茶专家团队一年半的策划，2017 年幻幻庵于涩谷开业，主要推广来自佐贺县嬉野市的茶叶。店铺选址在热闹且时髦的涩谷区，店面设计休闲而简约，幻幻庵团队也是希望通过这样的设计可以吸引不怎么喝茶的年轻客人进来。柜台上不展示任何茶叶，而是将每种茶水取一些放在小玻璃瓶中，摆在对应的茶叶铁盒前供人观赏与选择。泡茶时店家会用上一些稀奇的器具，例如烧杯，或者中国台湾生产的中式茶壶。再比如在冲泡浓郁的玉绿茶时，店家会将泡好的浓茶倒入一个小杯子里，不由得让人联想到意式浓缩咖啡。幻幻庵的经营理念就是：让踏进店里的人们以品一道茶为契机，从而进一步激发他们对于日本茶的兴趣。

㊋11:00—23:00 ㊄无 ㊡不固定 (址)涩谷区宇田川町 4-8 (交)JR 山手线或其他线路涩谷站步行 5 分钟 (网)www.gengenan.net

⑭ GEN GEN AN (Shibuya)

GEN GEN AN is a nihoncha specialty cafe, but it looks as if it is a bar with its walls painted in black with its English signboard. The muffled sound of African-American music streams out. The café was opened in 2017, by nihoncha specialists with an aim to increase distribution of nihoncha from Ureshino, Saga. You can taste terroir, which can be translated as the characteristic taste imparted to the land where the chaba is produced. Shibuya and its casual atmosphere were chosen to attract customers who do not normally enjoy nihoncha. The café wishes that their experiences here will make them interested in nihoncha. Instead of chaba samples, brewed tea in the bottles are displayed to help customers pick their teas. Dark and thick ryokucha brewed by Mushi-Sei Tamaryokucha Seihou, a unique method to brew tamaryokucha, is served in a tiny cup, as if it is an espresso. If your interests grow, then the café staff will respond to your questions on chaba or brewing methods.

(H) 11:00-23:00 (Friday-Sunday:11:00-24:00) (T) None (C) Irregular Closure Days (Ad) 4-8 Udagawa-cho, Shibuya-ku (Ac) 5-minute walk from Shibuya Station (JR Yamanote Line, etc.)
(U) www.gengenan.net/en

照片上为红色焙茶（450 日元）。幻幻庵出售的茶叶颜色都比较鲜艳，您可参考前台摆放着的茶叶样本再选择。

In the photo Beni Hojicha. (450 yen)
The uniqueness of nihoncha at GEN GEN AN is its colorfulness.
Brewing Sample is poured into a glass bottle similar to a medicine one. A great reference when you are ordering.

店名“幻幻庵”来自江户时代的“卖茶翁”建的最后一栋茶庵的名字。卖茶翁是江户时期有名的茶商，是他将煎茶推向了市场。罗马字及汉字的店铺标识是由法国设计师设计，这样的店标让外国游客也能一目了然，可以吸引多元的顾客光顾茶店。

The café name originates from the last place Baisaou created. Baisaou spread sencha during the Edo period. The logo with roma-ji and kan-ji is designed by a French designer. It is easy for foreigners to understand and leaves a strong impression.

照片为大杯绿茶（450 日元），在炎炎夏日，一口气喝上一杯，十分爽快。茶本来就是“能量饮料”，大口大口地喝下去，提神醒脑，让人神清气爽。除此之外，店家还提供柚子绿茶与柠檬草绿茶的冷饮。

If you are thirsty, then getting an L size (Large), cold brew nihoncha in a deep green color is recommended. (450 yen). Nihoncha is essentially an energy drink. so if you drink nihoncha in big gulps, you will feel energized. Yuzu-ryokucha and lemongrass-ryokucha are also available as iced drink.

店内可以买到古法制作的炒茶，以及无农药栽培的洋甘菊与柠檬草混合的焙茶。茶叶单买（50 克 1200 日元）、茶包（7 包 800 日元）。

Tea leaves are available as chaba (50g/1,200 yen) or tea bags (7 tea bags/800 yen).
All teas are carefully selected or blended: Simple Kamairicha, produced utilizing a traditional method; Organic Chamomile tea; and Hojicha with Lemongrass.

其他推荐
BEST CAFES and MORE

Chakas 涩谷店（涩谷）

精选茶叶冲泡出的美味茶饮搭配店主亲手制做的饭团。

(时) 9:00—18:00 (电) 03-6427-3059 (休) 不固定 (址) 涩谷区东 1-3-1 kaminito 大楼 1 层 (交) JR 山手线或其他线路涩谷站步行 10 分钟 (网) chakas.tokyo

CHAKAS (Shibuya)

Carefully selected nihoncha brewed with kyusu. Enjoy with hand-made onigiri.

(H) 9:00-18:00 (T) 03-6427-3059 (C) Irregular Closure Days (Ad) Kaminito Building 1st Floor, 1-3-1 Higashi Shibuya-ku, (Ac) 10-minute walk from Shibuya Station (JR Yamanote Line, etc.) (U) chakas.tokyo

---

Chakas 住吉店（住吉）

位于深受外国背包客喜爱的一间青旅内。

(时) 9:00—17:00 (电) 03-6666-4377 (休) 星期二 (址) 江东区扇桥 2-4-4 (交) 都营新宿线、地铁半藏门线住吉站步行 7 分钟 (网) chakas.tokyo

CHAKAS (Sumiyoshi)

Opened within the guesthouse for foreign tourists!

(H) 9:00-17:00 (T) 03-6666-4377 (C) Tuesday (Ad) 2-4-4 Ogibashi, Koto-ku (Ac) 7-minute walk from Sumiyoshi Station (Toei-Shinjuku Line or Tokyo Metro Hanzomon Line) (U) chakas.tokyo

---

Tea Press 东京原宿店（原宿）

在以“蓝色”为主题的店铺里沏上一壶好茶。

(时) 11:00—15:00 (电) 03-5786-3303 (休) 星期六、星期日、节假日 (址) 涩谷区千驮谷 3-7-6 (交) 地铁副都心线北参道站步行 3 分钟 (网) www.instagram.com/teapressharajuku

Tea Press - TOKYO Harajuku (Harajuku)

Enjoy nihoncha brewed with kyusu at an elegant space.

(H) 11:00-15:00 (T) 03-5786-3303 (C) Saturday, Sunday and Holidays (Ad) 3-7-6 Sendagaya, Shibuya-ku (Ac) 3-minute walk from Kita-Sando Station (Fukutoshin Line) (U) www.instagram.com/teapressharajuku

---

Luvond Tea（银座）

可以选择来自世界各地的茶品。

(时) 11:00—23:30 (电) 0570-001-432 (休) 以 Ginza Six 商场为准 (址) 中央区银座 6-10-1 Ginza Six 商场 6 楼 (交) 地铁银座站 A3 出口步行 2 分钟 (网) ginzagrand.jp

LUVOND TEA (Ginza)

Enjoy carefully selected teas from around the world at the counter.

(H) 11:00-23:30 (T) 0570-001-432 (C) Same hours as GINZA SIX’s operation hours (Ad) 6-10-1 Ginza, Chuo-ku (Ac) 2-minute walk from Ginza Station Exit A3 (Tokyo Metro). (U) ginzagrand.jp

第三章

习茶：
学习日本茶的种类与泡茶方法

LEARN ABOUT
NIHONCHA

让每个人都能将心仪的好茶泡出最佳的味道

一保堂茶铺是一家有近 300 年历史的日本茶专卖店，店名中的“一保”意为对茶道一心一意的专注，是这家百年老店代代相传的核心精神。一保堂总店位于京都府上京区的京都御所（又名“京都皇宫”）附近，而东京的分店则开在东京中心地区丸之内。两家店内都设有供客人歇脚的“嘉木”茶室，在茶室内客人不仅能喝到不同的茶品，店家还会指导客人泡茶方法，目的是让客人选购心仪的茶品回家之后也能泡出茶叶最佳的味道。虽然不同品种的茶泡法各不相同，但只要掌握基本的流程，泡茶并非难事。还请踏入一保堂挑选自己心仪的茶品，然后学习如何泡出一杯好茶吧。

To Brew Your Own Favorite Nihoncha

With a history going back nearly 300 years, Ippodo, which literally means preserve one, is a nihoncha specialty store. At the Kaboku Tearoom, adjacent to the main store in Kyoto near Kyoto Imperial Palace or its Marunouchi store in Tokyo, Ippodo offers a style of service in which customers learn how to brew nihoncha for themselves. This service style aims to teach customers how to brew nihoncha so they can brew their own cup of tea at home. The brewing method can vary depending on which varieties of nihoncha you choose, but it is not so difficult to bring out the best in all nihoncha as long as you know the basics. Find your favorite nihoncha at Ippodo and also learn how to brew it, so you can bring the best taste out of the tea by yourself at home.

ARTRIP ADVISER
艺术之旅顾问

一保堂茶铺
东京丸之内店
IPPODO TEA Tokyo
Marunouchi Store

一保茶茶铺是一家专注销售以京都宇治地区制法产出的“京都茗茶”的老店，总店在京都府。东京丸之内店是唯一一家与总店一样为门面店的分店，招牌显眼且方便客人的进出。丸之内店位于东京的中心地区，靠近东京站以及皇居等游客聚集的地方，因此有很多外国游客会到访店面。

IPPODO is a tea specialty store, headquarters in Kyoto with a long history. Its signature, high-quality Kyo Meicha is made with production method from Uji, Kyoto, called Uji Seiho. Conveniently located in the center of Tokyo near Tokyo Station and Tokyo Imperial Palace, Marunouchi store is the only street-front store of IPPODO and popular among foreign tourists.

⑮ 一保堂茶铺东京丸之内店（丸之内）

一保堂茶铺东京丸之内店于 2010 年开业，位于东京的“玄关口”丸之内地区，区内高层办公楼鳞次栉比，各大时尚品牌的专卖店错落有致，一间百年老店就这样坐落其中。虽然是一家新开业的门面店，但店面装饰与京都总店如出一辙，店门挂着标志性的厚重暖帘，暖帘上印染着“茶一保堂”的白色店名，展现出老店底蕴深厚而典雅的气质。在这家店里，客人可以通过购买茶叶、享受茶品以及学习泡茶来体验日本茶的魅力。店内分为三个区域：商品销售处，在这里选购茶叶前可以试饮；活动空间，店家会定期举办一些指导泡茶的活动；“嘉木”茶室，客人在这里点好一例茶后，可以在店家的指导下自己泡茶。由于茶叶的种类、泡茶水温等不同，泡出来的茶味各不相同，店家会为客人根据茶的口味搭配不同的和果子。店员们都是经验丰富的专业茶艺师，客人可以毫无负担地向他们咨询关于茶叶的问题，最终寻得最符合自己喜好的茶叶。

㊐11:00—19:00（最后点单时间 18:30）㊁03-6212-0202 ㊡年末年初及大厦休息日 ㊟千代田区丸之内 3-1-1 国际大厦 1 层 ㊋地铁有乐町站、日比谷站 D1 出口步行 5 分钟 ㊓www.ippodo-tea.co.jp

⑮ IPPODO TEA Tokyo Marunouchi Store
(Marunouchi)

IPPODO's Marunouchi store opened in 2010 on the spacious street where office buildings and famous brands' street-front stores are lined up. Marunouchi is known as Tokyo's gateway. Customers will be drawn to its elegant and aesthetic atmosphere with a profound noren, identical to its main store in Kyoto. Marunouchi store offers its customers core experiences of purchase, fun and learn at 3 respective spaces at the store, Retail Space, Event Space and the Kaboku Tearoom. Customers can learn about the intriguing joys nihoncha has to offer. At Kaboku Tearoom, you brew your own nihoncha after ordering your choice of chaba. With each chaba you choose, there is uniqueness in selecting the best temperature of the water to steep the tea or the kind of Japanese confectionary to be paired with. The store staff are knowledgeable, so feel free to ask questions and find your favorite.

(H) 11:00-19:00 (L.O.18:30) (T) 03-6212-0202 (C) End and Beginning of the Year. Building Closure Days (Ad) Marunouchi Naka-Dori Kokusai Building 1st Floor, 3-1-1 Marunouchi, Chiyoda-ku (Ac) 5-minute walk from Hibiya Station Exit D1 (Tokyo Metro Yurakucho-Line). (U) www.ippodo-tea.co.jp/en

一保堂
御銘茶

茶室给人一种安心感，北欧风格的椅子靠在原木色调的长桌旁，坐在靠内侧的位置可以看到吧台后方的店员点茶的样子。

Inside of tearoom, natural wood tables and comfortable chairs from Scandinavia create a welcoming and relaxing atmosphere. At the counter seats in the back of the tearoom, customers can view the process of matcha whisking.

日本茶的四个基本种类

4 BASIC TYPES OF NIHONCHA

玉露的特征是茶叶带有甘甜味。这是因为在茶树栽培过程中，当茶树刚刚冒出新叶芽时便被遮上帆布以避免阳光直射，这样的做法避免了茶叶里含的甘味成分（茶氨酸）接触阳光后转换为涩味（儿茶酸），因而形成了玉露温厚浓郁的气味以及独特的甘甜与清香。玉露还有一个特征是其茶汤色泽与其他茶品相比更为清亮。

Gyokuro tea leaves contain a sweet characteristic, because they are covered and shaded from the direct sunlight during the new leaves' growing period. In this cultivation process, theanine, the component contained in nihoncha tea leaves, is prevented from converting into catechin, the source of shibumi. The result is that gyokuro has a full-bodied taste with a distinctive sweetness and flavor. Its suishoku is also unique as it is clearer than other nihoncha.

天下一

TENKAICHI

"天下一"是一保堂茶铺所有的玉露茶中口味最醇厚的一个品种，保留了浓缩的茶叶精华，茶香缠绕舌尖，余味不尽。推荐用 60 摄氏度左右的热水冲泡。

Among Ippodo's gyokuro, it has the richest taste. All the essence from the tea is concentrated into Tenkaichi. You may want to roll the tea over your tongue to enjoy the lingering taste of this gyokuro. Recommended to brew Tenkaichi with the hot water at 60 degree Celsius.

芽茶

MECHA

芽茶是指在茶叶刚冒出茶芽时便收采的品种。茶叶细小呈圆球形状，并非如一般茶叶那般呈针状。茶水偏淡淡的苦涩，但茶叶上浓缩了甘味的精华，喝起来依旧口感香醇。

Nihoncha with the early leaf buds. Due to small leaves, chaba are not a needle-like shape, but rather rolled up in a small ball shape. Slight bitterness and nutrients are packed into Mecha and its rich taste is distinctive.

茎玉露

KUKI GYOKURO

茎玉露，顾名思义，是将茶叶的茎部采摘下来的茶品种。茎部的甘甜虽然没有叶片那般浓郁，但清淡的甜味也是很独特的。

This Kukicha is made of stems of gyokuro. The unique taste of this gyokuro, umami, is not as prominent as in leaves for stems. You will enjoy the refreshing clean taste with a hint of sweetness, which is unique to the stems.

粉茶

KONACHA

粉茶是指在玉露的制作工序中，将筛选茶叶后留下的茶粉集中起来的一个分类。只需将茶粉放进茶杯里，倒入开水即可快速品尝到玉露的风味。

Konacha is made up of powders of gyokuro leaves. As Konacha is in powder form, you can enjoy the taste of gyokuro by putting Konacha on chakoshi and quickly dipping the chakoshi through the hot water.

煎茶的特征是茶水的口感清爽。在种植过程中，茶叶充分接受阳光的洗礼，茶叶内的儿茶酸会增加，所以煎茶类的茶叶涩味会比玉露明显。煎茶的甘味与涩味恰到好处，适合大众的口味。一保堂茶铺供应有机栽培的煎茶。

Sencha is grown in full sunlight, which contains a large catechin compound, resulting in its distinctive shibumi. Its taste is smooth and refreshing. As it has a good level and balanced sweetness, Sencha is an all-around, all-occasion nihoncha. At Ippodo, there is also an organic option of sencha sold as well.

嘉木
KABOKU

嘉木是一保堂茶铺最具代表性的煎茶品种。这个品种用 80 摄氏度的热水冲泡最佳，甘味、涩味以及香味都会让人回味无穷。

Ippodo's signature sencha blend. It is best to brew with 80 degrees Celsius hot water, in order to maximize each characteristic taste of amami, shibumi and aroma.

茎煎茶

KUKI SENCHA

茎煎茶只选取质量上乘的煎茶茎部作为原料。这一品种不仅有煎茶本身的风味，还带有茶茎部凝聚的甘味以及独有的清香，用热水冲泡更能品尝出茶叶的甘醇本味。

Kuki Sencha is made of stems of only the highest-quality sencha. In addition to the unique taste of sencha, you can enjoy the rich sweetness and refreshing fragrance unique to stems. It tastes great with hot water.

粉茶

KONACHA

这是收集煎茶茶叶筛选后留下的粉末而来的一个品种，形状如同干粉状的青海苔。在泡法上，不用等着盖上茶壶盖让茶叶泡开，而是直接冲泡茶粉即可。

Konacha is made of small bits of sencha chaba. Shaped like aonori-dried green laver spice, konacha of sencha does not require the opening of leaves for brewing, so it is easy to brew with chakoshi.

番茶分为两种类型：将煎茶叶进行煎炒与烘焙后的茶为焙茶；将煎茶叶与煎炒过的玄米混合在一起的茶为玄米茶。种类繁多的番茶只需要注入热水便可饮用，味道清淡，做法简单，是日常生活中不可或缺的茶饮。

Hojicha is made from sencha tea leaves which are grown large and, roasted. Genmaicha is made with nihoncha, and roasted genmai-brown rice, added. Bancha comes in many varieties and is easy-to-brew with hot water. Bancha is perfect as an everyday tea and can be quickly brewed to enjoy.

极上焙茶
GOKUJO HOJICHA

为了避免茶叶被炒焦，这一品种的番茶一般会用小火煎炒。在煎炒过程中茶叶中的咖啡因会渐渐减少，因此饮用这类茶不需要顾虑太多咖啡因摄取过量等问题，适合需要补充水分时饮用。

Gokujyo Hojicha is roasted slowly with care being taken not to burn chaba. Caffeine is reduced during the process, so you can drink Hojicha without worrying about the caffeine intakes. Hojicha is perfect for hydration.

茎焙茶

KUKI HOJICHA

这一品种的焙茶主要选取的是大叶片煎茶的茎部，长时间煎炒至散发香气。用茎焙茶泡出的茶水比焙茶茶水的颜色更浓，自带一些甜味。

Kuki Hojicha is made with sencha stems which are grown longer than ideal length for nihoncha. The roasting process for stems takes longer than leaves. The suishoku of this Kuki Hojicha is darker and it has a sweetness unique to stems.

极上玄米茶

GOKUJO GENMAICHA

将煎炒好的玄米混入煎炒好的茶叶里，茶叶比玄米的分量更多，这样一来，泡出的茶水不仅有煎茶的甘涩，还有玄米的清香。

Gokujo Genmaicha is made with sencha tea leaves which are grown large, with roasted genmai added. The ratio of chaba is larger so you will feel the refreshing shibumi while enjoying the rich flavor of genmai.

抹茶是一种绿茶粉。跟玉露一样，茶叶在采摘前会用帆布遮盖起来，让甜味存留在叶片上。采摘下来的茶叶称为碾茶，碾茶经过蒸制干燥、用石磨碾磨成粉状后便是人们熟悉的抹茶。抹茶的特点是色泽呈鲜绿色，茶粉中含有茶叶的营养精华。抹茶的点茶手法是需要用茶筅将水与茶粉来回搅拌，打出茶沫则可享用。

As the same cultivation process for gyokuro, matcha is based on tencha tea leaves, which are grown under covers and in shade. Tencha tea leaves contain umami. Matcha is made with tencha grinded with a stonemill. A bright green color is its characteristic and you can intake chaba's umami as well as all nutrients. Matcha is unique in that it is whisked with chasen, a bamboo whisk.

云门之昔

UMMON-NO-MUKASHI

一保堂茶铺共提供 10 个品种的抹茶，会根据茶道的流派表千家与里千家的技法与礼仪为不同种类的抹茶命名。调制一人份的抹茶，浓茶需要茶粉 4 克（约 2 小勺），薄茶需要茶粉 2 克（约 1 小勺）。

Ippodo's Matcha has 10 levels of richness. Macha names vary, depending on which matcha preparations style is used as there are urasenke or omotesanke disciplines in sado, a traditional Japanese tea culture. For koicha, approximately 4g (2 small spoons) and for usucha, approximately 2g (1 small spoon) are used for 1 serving.

在一保堂茶铺选购质量上乘的好茶
Get High-Quality Nihoncha at IPPODO TEA

在茶铺可选购不同品牌的玉露与煎茶（有 100 克装与 200 克装），包装的颜色因茶名的不同而有所差异，各种颜色带来的不同感觉方便客人轻易地找到属于他们的那一款茶。

Depending on the blend, the packaging colors vary for gyokuro and sencha products (100g or 200g). Ippodo hopes that customers will find their own favorite teas, and that they will identify their favorite nihoncha by remembering the color of the package.

图中的茶罐只用来装玉露与煎茶，罐子上橘黄色的标签是明治时代流行的设计，标签上绿色茶壶的图案中间写着这款茶的品名。茶罐有大中小三种型号可以选择，放入专门设计的礼品盒里，不失为一件送礼佳品。

The vivid orange colored label is a modern design from the Meiji period, exclusively used for gyokuro and sencha. Chamei, the chaba name, is written on the green tea caddy. Tea caddies are available in large, medium and small sizes. A specialty gift box is also available.

图中这款礼盒里有一红一黑的特制茶罐，里边的茶叶是玉露与煎茶，桐木制成的礼盒盖子上印着一保堂茶铺的店标，显得高雅而大气。

In a wooden box made of kiri, meaning paulownia wood in Japanese, with the store name branded on it, gyokuro and sencha are placed in profound, lacquered tin (special chazutsu) in red and black for an elegant, sophisticated gift.

一保堂茶铺传授的美味泡茶法

HOW TO MAKE GOOD NIHONCHA, IPPODO TEA STYLE

在一保堂的“嘉木”茶室里，只要客人点上一壶茶，店员会送来热水壶、茶壶、茶叶、茶杯与时钟，手把手地教客人泡出一壶好茶。客人只要掌握了茶叶的分量、泡茶的水温、浸泡的时间等基本要点之后就可以尝试自己在家冲泡美味的好茶。在这一章节，一保堂将会简要介绍煎茶的泡法。

At the Kaboku Tearoom, when you order nihoncha, a pot with hot water, yunomi and a clock are brought to you. You can learn the brewing method and try to make your own cup of nihoncha. As long as you are attentive about the amount of tea leaves, the temperature of the water and the steeping time, you will be able to enjoy a good cup of nihoncha at home just like the one you have at the Kaboku Tearoom. Here, we'll introduce the brewing method of sencha.

① 放入茶叶

使用茶勺舀上 2 勺茶叶放入茶壶，一保堂的煎茶一般推荐 10 克（约两大勺）的量。

① PLACE CHABA IN KYUSU

For Ippodo's sencha, 10g of tea leaves (2 large spoons) should be used.
Prepare kyusu, and place chaba for the number of people you are serving for in kyusu with chasaji, a teaspoon.

② 将温度合适的热水注入茶壶

泡煎茶最合适的水温为 80 摄氏度。将烧开的热水先倒入专用的冷却茶碗内，一般来说热水温度会下降 10 摄氏度左右，还可以给茶碗保温。之后静置一分钟。同时也可以观察水蒸气的状态。

② PLACE YUSAMASHI WATER IN KYUSU

The ideal water temperature for sencha is 80 degrees Celsius. When we pour hot water in a cup, the water temperature becomes 10 degrees cooler. Pour boiled water in yunomi to cool it down to the ideal temperature and keep yunomi warmed. Wait for 1 minute. Keep an eye on how the steam is coming out.

③ **泡开茶叶**

将温度适宜的热水倒入茶壶，等待约 1 分钟让茶叶浸水泡开，推荐看着时钟或秒表来控制时间。最好让茶壶静置，如果期间摇晃茶壶的话，茶叶有可能会产生过多的涩味。

③ STEEP AND EXTRACT

When the hot water is at the appropriate temperature, pour hot water in kyusu. Wait 1 minute until chaba expands and completes extraction. It is good to use a clock or timer. Avoid moving kyusu because it may extract additional, unnecessary shibumi. Wait for the chaba to open up naturally.

④ **倒茶入碗**

由于茶碗还残留着冷却开水时的温度，将泡好的茶水一滴不漏地倒入碗里，一杯温暖身心的茗茶便做好了。每个品种的茶呈色各不相同，图中为一保堂的“嘉木”煎茶，茶水颜色为“山吹色”，即处于黄色与橘黄色之间的颜色。

④ POUR NIHONCHA INTO THE CUP

If you pour nihoncha into yunomi where you kept the hot water to cool down, you can drink your nihoncha warm. Suishoku(tea color) is different depending on each nihoncha. This sencha is Kaboku and its suishoku is yamabukiiro. Make sure you pour everything into the cup until the last drop.

⑤ **凭客人喜好可加水**

第一道茶之后，客人就可以根据个人喜好泡第二道、第三道茶……推荐每次在倒入热水之前都保持茶盖打开的状态，这样茶叶就不会过度浸泡。茶叶此时已经全被泡开了，从第二道茶开始只需直接倒入热水，无须静置便可直接倒茶饮用。

⑤ REFILL, IF YOU WOULD LIKE

You can enjoy a second, then a third cup of tea, but until you brew again, you need to keep the lid of the kyusu off so that the chaba is not over-steeped. From the second cup, chaba is already opened out, so pour tea into your cup immediately after you pour hot water into the kyusu.

可在一保堂茶铺享用的四类茶品

NIHONCHA MENU AT IPPODO TEA

玉露

使用60摄氏度左右的热水慢慢泡出来的玉露，有着浓厚的回甘。一保堂茶铺可提供的玉露茶品有推荐用60摄氏度左右的热水冲泡的“天下一”与“甘露”，以及推荐用开水冲泡的“万德”。上图为“天下一”玉露（2400日元）。由于低温慢泡，茶的甘甜浓到一定程度时甚至会带上“出汁”（鲜汁汤）的味道。饮一口浓茶，再搭配京都特产大德寺纳豆制成的小点心，别有一番滋味。

GYOKURO

Gyokuro is much richer in umami. Particularly for the full-bodied gyokuro, tea leaves are steeped slowly with the hot water at a temperature of around 60 degrees Celsius to extract umami. Tenkaichi and Kanro are brewed at 60 degree Celsius. Mantoku is brewed with hot water. The photo shows Tenkaichi (2,400 yen). Rich in umami and smooth in texture. You would love the cup with a unique wagashi made from Kyoto's daitokuji-natto.

煎茶

泡煎茶时，一般使用 80 摄氏度的热水便能泡出茶的甘甜、苦涩与醇香之味。对于一些回甘更浓的煎茶品种，可以倒入更高温的热水把香气“熏”出来。上图为“嘉木”(1600 日元)，最佳的泡茶水温是 80 摄氏度，可以将开水在茶碗里冷却。“嘉木”煎茶的茶水颜色介于黄色与橘黄之间，若冲泡出的茶水呈亮黄色，则代表茶叶品质较高。搭配用山药包裹红豆馅的和果子，着实美味。

SENCHA

When brewed at 80 degrees Celsius, sencha can offer a balanced taste, while all amami, umami, shibumi, nigami and aroma components are harmoniously balanced. If your sencha is strong in shibumi, then use hot water at a higher temperature to bring out the aroma. The photo shows Kaboku (1,600 yen). Cool down hot water to 80 degrees Celsius to brew. Bright golden yellow tea color is a sign of good taste. Paired with Joyo bun.

焙茶

焙茶的魅力就在于茶叶经过煎炒后散发的独特香气，以及清淡的口感。注入热水的瞬间，煎炒过的茶叶，其独特香气当下散开。一保堂茶室提供的极品焙茶（1000日元），呈色为红褐色。由于焙茶的口味偏淡，推荐搭配一些比较甜的和果子。上图的和果子，就是一款包裹了满满红豆馅的糯米糕。

HOJICHA

Hojicha is roasted nihoncha, rich in flavor and aroma while offering a lighter taste. As soon as hot water is poured over the leaves, a distinct rich flavor shimmers out. At the café, you can enjoy Gokujyo Hojicha (1,000 yen). Tea color of Gokujyo Hojicha is clear, reddish brown. It has a rich, roasted flavor with lighter taste, so it goes perfect with rich-tasting sweets. In the photo is a heavier Japanese confectionary which is an covered with mochi.

抹茶

一保堂提供三个品牌的抹茶，会根据客人是选择浓茶还是薄茶选上相宜的一款。在日常生活中已经很少有机会能喝到用茶筅打出来的正宗的浓茶了，因而一保堂的抹茶也显得尤为珍贵。在店里，选择了浓茶之后，客人还可以要求店员加水打成薄茶。图中为“云门之昔”(2200 日元)，推荐搭配口感顺滑、带有醇厚甘甜味的生果子一起享用。

MATCHA

3 varieties of matcha are available. Choose either koicha (thicker matcha) or usucha (more common, thinner matcha) preparation methods. Koicha is not commonly served so it is a rare opportunity to try it out. Koicha, can be whisked as usucha, later. Koicha, Ummon-no-Mukashi (2,200 yen). Omogashi (moist wagashi), delivers smooth and elegant sweetness when it is paired with matcha.

日本茶小知识

茶的营养成分

茶叶中包含碳水化合物、蛋白质、脂质、维生素、矿物质等营养成分，还有茶叶特有的儿茶酸、咖啡因以及茶氨酸等成分。营养成分大多不溶于水，因此在喝茶的时候，儿茶酸、咖啡因、茶氨酸以及维生素C等可溶于水的物质会被摄入体内，对人体健康有很大帮助。

茶的功效

茶叶中包含的最主要成分儿茶酸具有抗氧化的功效，能帮助去除人体内的有害活性氧。儿茶酸及其他化合物在控制高血压、降低血液中的胆固醇、预防肥胖、预防癌症等方面都具有一定的辅助功效，并已得到临床验证。饮茶确实能维持身体健康。

茶的保管方法

茶叶中包含的化合物在接触空气后易氧化变质，且湿度越高氧化速度越快，因此在保存茶叶的时候最讲究环境的干燥性。最好将茶叶放入密封的容器内，远离明火与阳光直射，并将其放在阴凉处保管。未开封的茶叶包装要套上一个塑料袋后放入冰箱保存，最长不超过半年；而开封过的茶叶放置时间不要超过1个月。

茶叶的特殊用法

放置时间较长的茶叶可以在加热翻炒后制成焙茶，或者可以放进香炉里当作线香使用。泡过的茶叶可以放进调料锅里作为烹煮食物的佐料，这样炖出来的料理不仅有茶的香味，还富有营养。干燥的茶叶甚至可以用来做除臭剂，因为茶叶包含的成分具有抗菌及除臭的作用。

[资料来源：世界绿茶协会]

TRIVIA ON NIHONCHA

NUTRITION OF NIHONCHA

Nihoncha tea leaves contain nutrients such as carbon hydrate, protein, lipid, vitamin and minerals, in addition to compounds such as catechin, caffeine and theanine. Most nutrients are not water-soluble. So, when you drink the cup of nihoncha, it contains catechin, caffeine, theanine and vitamin C, and you can benefit from these water-soluble compounds.

BENEFITS OF NIHONCHA

Catechin, the main compound and other compounds in nihoncha, has anti-oxidative benefits. They help to remove reactive oxygen species (ROS) which are harmful to our body. Many studies confirm that catechin and other compounds help with maintaining the healthy level of blood pressure and blood cholesterol as well as with preventing obesity and cancers. Drinking nihoncha can promote the maintenance of overall health.

THE STORAGE METHOD OF NIHONCHA

Compounds in nihoncha can deteriorate due to oxidation. Oxidation can be accelerated by the moisture, so keep it in a dry place. Store nihoncha in an airtight container and keep it in a cool place, away from fire or sunlight. If it is unopened, then keep nihoncha in the original packaging, covered with another plastic bag, and store in the refrigerator for 6 months after the production date. After opening, it is best to consume the tea within 1 month.

INTERESTING WAYS TO UTILIZE TEA LEAVES

If nihoncha becomes old or moist, you can roast it. Make tasty homemade hojicha. Enjoy the aroma on chakou ro, an item to burn chaba on for their aroma. Chagara, leftover tea leaves after tea is brewed, can be made into dishes by adding sauce or spices, for example to make tsukudani, Japanese dish with ingredients simmered in soy sauce and mirin. Absorb all nutrients from chaba. Dried chaba can be used as deodorant, as nihoncha has antibacterial and deodorant benefits.

(Credit: World Green Tea Association)

第四章

买茶：
逛遍日本茶好店

PURCHASE
NIHONCHA

在品类繁多的茶店内选茶，享受与店员的交流

奥斯卡·布莱克鲁来自瑞典，学生时代的他第一次在家乡的红茶店喝到日本茶时，对于日本茶的印象仅是“苦涩而难喝”。待他来到日本，真正喝到店家用茶壶泡出的茶后，才首次感受到地道日本茶的美味。奥斯卡说：“在日本以外的国家，茶是一种健康饮料，而茶在日本是一种嗜好，是一种可以细细品味的日常饮品。不仅如此，在日本可以选择的茶品数不胜数。”奥斯卡的足迹遍布全日本的茶咖啡店和茶铺，甚至各个茶园，从茶叶的种植开始学习，身体力行去发掘深层次的日本茶魅力。“若想找到适合自己口味的茶，最好的方式是到茶品繁多的茶店内直接与店员沟通，让店员可以根据个人喜好提出建议。”

Select your own chaba at the store which has a wide selection of nihoncha, while enjoying conversation with store staff.

Oscar Brekell, who works in Japan as a Nihoncha Instructor, grew up in Malmo, Sweden. He first encountered nihoncha when he was in high school there. “It was bitter and not good,” he smiles. Later, while in Japan, he had a cup of nihoncha properly brewed in kyusu for the first time. He experienced the cup with his heart and felt it was a truly beautiful taste. “Outside of Japan, the image of nihoncha is that of a health drink. In Japan, on the other hand, nihoncha is considered as shikohin, meaning luxury daily items. There are much wider selections of chaba varieties in Japan.” Since then, he visited nihoncha cafés, retailers of chaba and producers at their nihoncha farms. “In order to find your own, favorite chaba, it is best to go to the nihoncha store first. Pick the store which has a wide selection of nihoncha varieties, and then ask questions to the personnel at the store.”

ARTRIP ADVISER
艺术之旅顾问

奥斯卡·布莱克鲁
Oscar Bre kell

出生于瑞典的日本茶艺师。现在日本茶出口促进协会工作，同在日本及日本以外的国家做各类关于日本茶的演讲。著有《金发碧眼的日本茶传教士奥斯卡：我所爱的日本茶》一书。

Born in Sweden. Nihoncha Instructor. While working at the Japan Tea Export Council, he holds nihoncha seminars both in and outside of Japan. His first book, *Bokuga Koishita Nihoncha no Koto Aoimeno Nihoncha Dendoshi Oscar - Nihoncha that I have fallen in love with*, was just recently published.

中村家世代耕耘茶田有将近 100 年的历史，所拥有的茶田位于静冈县藤枝市的高海拔地区。中村茶生活馆的茶叶就是在这片茶园里种植的。茶园从 1979 年开始采用减农药栽培方法，1983 年改用无农药栽培方法并延续至今。

Tea farm owned by Nakamura Family in Fujieda, Shizuoka. The family has an approximately 100-year history in cultivating nihoncha at this farm. Nihoncha sold at NAKAMURA TEA LIFE STORE (P98~) are cultivated here. In 1979, the Nakamura family started reduced-pesticides cultivation. Since 1983, the family has been continuing the pesticides-free cultivation of nihoncha till this day.

⑯ 中村茶生活馆（藏前）

中村茶生活馆于 2015 年开业，茶店只提供采自静冈县藤枝市中村家茶园的茶叶，中村家于大正八年（1919 年）开业，至今也有百年历史了。中村茶生活馆的店主西形出生于藤枝市，他与中村家茶园的经营者从小便是好友。西形成长于盛产茶叶的静冈县，饮茶对于他来说就是个生活习惯。他初到东京时，惊讶地发现东京竟然鲜有可以泡茶的地方，从而萌生了要培养东京年轻人泡茶饮茶兴趣的想法，于是与友人一起创立了中村茶生活馆。店内使用的茶叶全是无农药栽培，而且由于种植区域的不同，茶叶的外形与口感还各有特征。由于中村家茶园有好几处种植区域，为方便区分，商品包装上会详细地标注“No.1”“No.2”或“No.3”来表示采摘的茶园。包装的设计由本职工作为设计师的西形店长亲自包办，包装上还标注了茶园所处地点的经纬度，简约而富有设计感。不同茶园产出的茶叶品种有煎茶、玄米茶、焙茶、茎茶以及抹茶。若客人实在犹豫不知该如何选择，可在店内试饮后再做决定。喝上一杯店员细致泡出的茶，感受一下用茶壶才能泡出的那份趣味。

㊐12:00—19:00 ㊁03-5843-8744 ㊡星期一 ㊟台东区藏前 4-20-4 ㊋地铁藏前站步行 5 分钟
㊙www.tea-nakamura.com

⑯ NAKAMURA TEA LIFE STORE (Kuramae)

Opened in 2015 in Kuramae, NAKAMURA TEA LIFE STORE only uses chaba from the reputable chaen (tea farm) , owned by Nakamurake, which was established in 1919 in Fujieda, Shizuoka. The owner, Mr. Nishikata is from Fujieda, and grew up together with the owner of Nakamurake. Being raised in an environment where nihoncha was a part of his daily life, he was surprised that in Tokyo, people often got curious of brewing the tea with kyusu. He hoped to get younger generations interested in nihoncha and so opened the store with friends. Chaba brewed at the café is cultivated with no pesticides. On the package label, chaen's location numbers, No1., No.2, or No.3, are listed with product information. Mr. Nishikata designed eye-catching packages as he is also a designer. Sencha from each chaen, genmaicha, hojicha, kukicha, funmatsu-matcha (matcha powder) are available. Ask store staff for questions. Tea samples are available. You will surely become interested in brewing nihoncha with kyusu.

(H) 12:00-19:00 (T) 03-5843-8744 (C) Monday (Ad) 4-20-4 Kuramae, Taito-ku (Ac) 5-minute walk from Kuramae Station (Tokyo Metro) (U) www.tea-nakamura.com

使用茶壶泡茶时，不同的水温以及不同的泡茶方法，会改变茶水的口感。能够感受到茶水在甘味、涩味上的细微改变，就如同发现生活中的小确幸。店长希望客人能够将泡茶这件事融入到生活中去，因此将店铺取名为“茶生活馆”。

You will discover when you brew nihoncha in the kyusu, that the degree of shibumi or amami changes depending on the temperature or the methods of brewing the nihoncha. In addition, you will feel relaxed by the aroma of the tea. The owner's wish of incorporating nihoncha into the individual's daily life, is in the name of the shop.

中村茶生活馆位于偏离主干道的一个分岔路口，在一座红砖外墙的矮层建筑内，店面的入口处挂着深蓝色的暖帘。这附近有很多质量上乘的手工作坊，来往的客人偏好进入店里跟店员交流，去了解产品的个性与背后的故事。这也是店主选择将生活馆开在藏前的原因。

The store is located on the street one block away from the main street. It is a brick building with a landmark navy-colored noren. In the neighborhood, there are many shops where they sell items with high-quality craftsmanship. Customers tend to purchase items after asking questions to the store staff. This was one major reason why NAKAMURA TEA LIFE STORE opened in this neighborhood.

店铺所在地原本是一家瓷砖工厂，西形店长与朋友亲自设计并装修，保留了老厂房的粗糙地面以及木制的置物架。店内时刻散发着的茶香将人带回了故乡，仿佛回到了外婆家，让人徜徉在记忆中的那段旧时光里。

It was a tile factory. Nishikata and his friends renovated the building while retaining the original earth floor and wooden shelves. The melancholic atmosphere and aroma of nihoncha will remind guests of grandma's house or home.

墙壁上挂着一块黑板，黑板上用粉笔详细绘制了泡茶方法和要点。这个板绘是由以为咖啡店制作独特板绘而闻名的艺术团队“粉笔男孩”（Chalk Boy）专门创作的。

On the blackboard above inside of the store, brewing methods and points are written in chalk. It is a work by Chalk Boy, famous for its unique artworks for cafés.

店内时刻被淡淡的茶香包围着。店家使用的茶香炉可以在网上买到，是一件增添日常生活乐趣的小物件。

There is chakoro placed inside of the store, so a light aroma of nihoncha will welcome you at the store. This chakoro is easy to purchase online, easy to use and can also be used at home.

上图的木箱是中村茶生活馆在大约 80 年前用来搬运与保管茶叶的工具，现在作为收纳与装饰放在高架子上，与店内复古的装修氛围相当契合。

The old wooden boxes used to transport or store nihoncha about 80 years ago are utilized for storage and displays. They match perfectly with the retro-atmosphere of the store.

茶袋上的标签以年轻一代，尤其是男性的时尚品位作为设计的出发点，具有简约时尚感，适合放在生活空间内。除了茶袋包装，从 No.1 到 No.3 的茶园产的煎茶、玄米茶以及混合了抹茶的玄米茶都有罐装，茶叶罐由浅草的匠人设计。

These labels are designed in consideration of younger generations, especially men. Also, they were designed in a way that makes them fit and match into daily living space. Sencha from No.1~3 chaen, genmaicha and genmaicha with matcha are also available in chazutsu. Each chazutsu crafted by craftsmen in Asakusa.

⑰ 乐山（神乐坂）

乐山茶店位于神乐坂地区，店内主要销售来自静冈县挂川市的茶叶，茶叶种类在 100 种以上。开业 50 多年以来访客络绎不绝，既有住在附近的街坊邻居，又有来自静冈县的客人，外国观光客也会光临。店内销售的茶叶从煎茶、玄米茶、焙茶到茎茶，应有尽有。除了茶叶，还可以选择不同的包装，有日常使用的袋装或者罐装；若要将茶叶作为伴手礼送给亲朋好友，还可以选择礼盒或者桐木盒装。店内装修古典而大气，陈列柜台内工整排列着各类品牌的茶叶，乍看之下，这样的陈列装潢似乎会给人一些“距离感”，但摆放出来的商品中不乏简易茶包以及调味茶，以尽量满足客人不同层次的需求，让客人可以尽情选购。若客人不知从何选择，可以先试饮并听取店员给出的意见再做决定。每个带“5”的工作日以及每周六，店家会在门口现炒焙茶，肆意的茶香让人不禁驻足，并踏入这个琳琅满目的茶世界。

㊐ 9:00—20:00（星期六 9:30—20:00，星期日、节假日 10:00—18:00）㊊ 03-3260-3401 ㊡ 无
㊄ 新宿区神乐坂 4-3 ㊋ JR 饭田桥站西口或地铁饭田桥 B3 出口步行 5 分钟
㊙ www.kakuzan.co.jp

⑰ Rakuzan (Kagurazaka)

Rakuzan opened in Kagurazaka 50 years ago. The store mainly offers nihoncha from Kakegawa, Shizuoka, and the store is often crowded. Varieties of nihoncha are prolific and almost reach 100 in total. Sencha, genmaicha, hojicha and kukicha and other selected teas are available in several packaging options, a bag, a tin, a gift box and a high-quality kiri gift box. You can find nihoncha for your home and for elegant gift items. The profound storefront and all the items lined up in the showcase may overwhelm you, but they sell casual items, such as nihoncha in an easy-to-brew individual tea bags or flavored tea items, so don't feel hesitant to enter the store. When you go in, store staff will serve you with a sample nihoncha. Find your favorite tea by asking advice from staff. Every month, on the weekday which includes the number 5 and every Saturday, the store will be filled with roasting aroma of hojicha as hojicha will be roasted in front of the store.

Ⓗ 9:00-20:00 (Saturday:9:30 ~ ,Sunday and Holidays:10:00-18:00) Ⓣ 03-3260-3401 Ⓒ None
(Ad) 4-3 Kagurazaka, Shinjuku-ku (Ac) 5-minute walk from Iidabashi Station, Exit B3 (Tokyo Metro) or Iidabashi Station West Exit (JR Line) Ⓤ www.rakuzan.co.jp

⑱ 辻利银座店（银座）

辻利银座店是京都宇治的老茶铺辻利的第一家东京分店，辻利品牌在日本可谓家喻户晓，由初代店长辻利右卫门创立，已有超过 150 年的历史。由辻利右卫门开创的玉露制法延用至今，使用古法研磨出的抹茶制成的点心也是招牌商品。银座店内销售来自高级茶产地京都宇治的各类茶叶，客人不仅可选购如“辻利大门”（107 页左侧；100 克 3 800 日元）这样的高级玉露，还有煎茶、焙茶、玄米茶以及抹茶可供选择。店内的人气糕点是洒满抹茶粉、口感香醇的“京都浓茶蛋糕卷”（2 000 日元），以及用抹茶与白巧克力低温烘焙而成的“京都浓茶巧克力酱糜”（2 500 日元），这款巧克力酱糜仅在银座店出售。在卖场旁边设有外卖窗口，菜单上供应抹茶粉配牛奶调制的“京都拿铁”、宇治煎茶以及银座店限定的“辻利浓茶”甜筒等。

㊐ 10:30—20:30 ㊀ 03-6263-9988 ㊡ 根据 GINZA SIX 商场的营业时间 ㊟ 中央区银座 6-10-1 GINZA SIX 商场 B2 层 ㊋ 地铁银座站 A3 出口步行 2 分钟 ㊓ www.tsujiri.jp

⑱ TSUJIRI Ginza (Ginza)

TSUJIRI with a main store in Uji, Kyoto and over 150 years of history, opened its first store in Tokyo. It is famous for its matcha, carefully ground on a stone mill, and sweets. Another signature item is gyokuro. Its production methodology is said to be established by its founder, Tsujiri Uemon. Gyokuro with soft aroma and umami, come in several brands, such as Daimon (100g/3,800 yen). Other nihoncha, such as sencha, hojicha, genmaicha and matcha, are available, all of which have characteristics of the climate and natural environment of Uji, the land known for the sophisticated, high-quality nihoncha. Popular sweets include: Kyo-Koicha Baumu (2,000 yen) with deep matcha taste and the melt-in-the-mouth texture; and Kyo-Koicha Terrine (2,500 yen) which is the sweets made with matcha and white chocolates. Kyo-Koicha Terrine is sold exclusively at the Ginza store. At a café stand, enjoy drink items such as Kyo-Latte or Uji Sencha, and Ginza store's exclusive item, TSUJITI Koicha, soft serve ice cream.

(H) 10:30-20:30 (T) 03-6263-9988 (C) Same hours as Ginza Six's operation hours (Ad) 6-10-1 Ginza Chuo-ku (Ac) 2-minute walk from Ginza Station, Exit A3 (Tokyo Metro) (U) www.tsujiri.jp

⓳ 石崎园（门前仲町）

门前仲町在江户时代是工商阶层的生活区，热闹繁华的街道附近有深川不动堂以及富冈八幡宫。石崎园茶铺创立于明治二十年（1887 年），位于车站附近的深川仲町商业街，是一家日本茶专卖店。店内陈列着有年代感的茶壶茶杯，还摆放着许多古旧的茶箱，时光仿佛在这里停滞了。与时光一道停留的还有那供应了 130 年、味道都没有变化的特色茗茶，最近店家将特色茶注册了“门仲茗茶”的商标，商标的图案是一只鸽子，以富冈八幡宫的神使信鸽为设计灵感。各类茶叶摆放在店门口的陈列台上，方便客人路过选购。石崎园的茶叶作为一款带有老街特色的伴手礼，深受过路的年轻人以及外国游客的欢迎。供应的茶品有玉露、煎茶以及深蒸茶等，其中煎茶还分为“碧”“冴”“雅”“香”“选”“摘”等品牌。当前的店主石崎是第四代店主，他持有茶匠（专业茶叶经营者）资格证，当客人在选择茶品上有任何犹豫或不解时都可以向他请教。

㊐ 10:00—19:00 ㊀ 03-3643-4188 ㊡ 星期日 ㊅ 江东区富冈 1-10-10 ㊂ 地铁东西线门前仲町站 1 号出口步行 1 分钟 ㊃ 无

⑲ ISHIZAKIEN (Monzennakachō)

Monzen-Nakacho offers the nostalgia of shitamachi, which literally means old downtown Tokyo, represented by shitamachi places, such as Fukagawafudoson Temple and Tomiokahachimangu Shrine. A commercial district strip, Fukagawa Nakacho Dori Shotengai, runs through in front of the train station. ISHIZAKIEN, a specialty nihoncha store, was established here in 1887. Profound chaki items are on display, and traditional chabako are lined up inside, very appropriate for the teashop of long standing. The taste of its signature brand, Monnaka Meicha, has been unchanged for the last 130 years. Monnaka Meicha is packed in a unique package with a motif design based on a dove, a common image used at Tomiokahachimangu Shrine. Monnnaka Meicha is popular among young visitors or foreign tourists as souvenir items from shitamachi. Varieties include gyokuro, sencha and fukamushicha. Among sencha, there are several brands available, such as Ao, Sae, Ko, Sen and Teki. Ask Mr. Ishizaki, the 4th generation storeowner, who is Chasho (Tea Business Management Specialist), your questions.

Ⓗ 10:00-19:00 Ⓣ 03-3643-4188 Ⓒ Sundays (Ad) 1-10-10 Tomioka, Koto-ku (Ac) 1-minute walk from Monzen-Nakacho Station (Tokyo Metro Tozai Line) Ⓤ none

⑳ 鱼河岸茗茶银座店（银座）

鱼河岸茗茶茶铺创立于 1931 年，总店位于以海鲜批发闻名的筑地市场，在东京开有多处分店。银座分店被装修得像是一家时髦的无座小酒馆，只要一踏入店里，穿着白衬衫黑背心的店员便会为客人端上一杯试饮茶，这种和洋折中的做法在茶店中实在是新颖。茶叶采用茶铺原创的“香薰制法”制成，因此有着口味上乘的苦涩与清香，茶水呈金色且透明，颇具特色。茶叶包装是由著名的画家和插画家设计，例如和田诚、宇野亚喜良、滩本唯人、南伸坊以及伊藤方也。包装的设计带有体现店铺理念的温馨感和幽默感，既适合带回家摆放，也可以作为送礼佳品。其中“茶茶 CHA”（800 日元）这款茶包组合最适合作为纪念品，组合内包含 3 种类型的煎茶、1 种釜煎茶以及 1 种焙茶，冲泡即可饮用。店内还展示及销售茶壶等茶具。

(时) 11:00—18:00 (电) 03-3571-1211 (休) 星期一 (址) 中央区银座 5-5-6 (交) 地铁银座站 B3 出口步行 1 分钟 (网) www.uogashi-meicha.co.jp

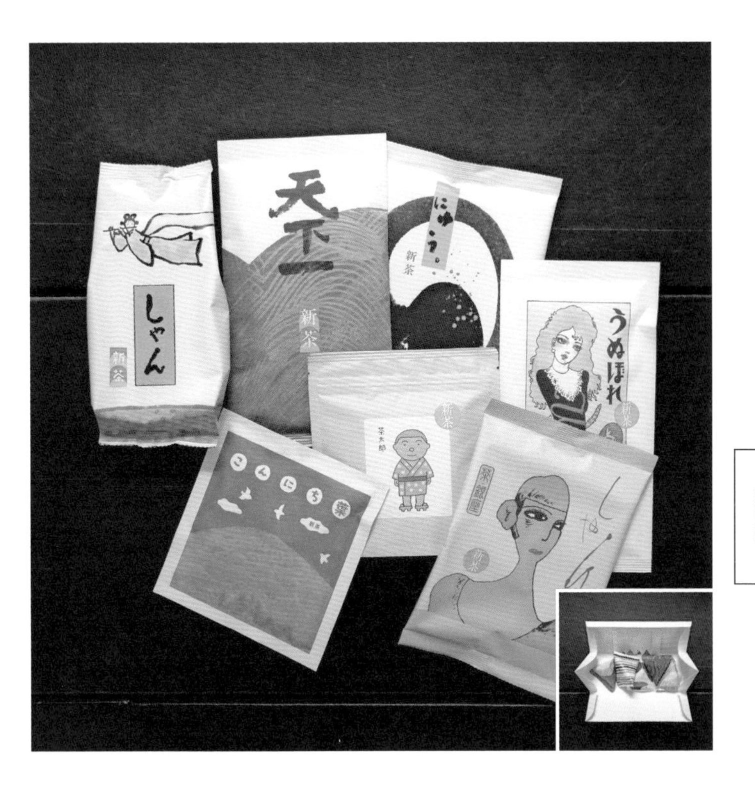

⑳ Uogashi-Meicha CHA・GINZA (Ginza)

At a glance, it is a stylish space that looks as if it is a modern bar. From the counter situated at the left side of the store, staff dressed in a white shirt with a black vest will swiftly serve you a sample cup of nihoncha. Established in 1931, Uogashi-Meicha's main store is located in Tsukiji Jogai Market. You will be served with a sample tea when you enter the store. Enjoy the balanced, combined taste of niga-shibumi and a good aroma in addition to its "clear, golden" color. Uogashi-Meicha's tea utilizes its original nihoncha production method, "Kaori-Mushi Seihou, literally means Aroma Steamed Production Method." Illustrations or calligraphy designed by well-known artists, such as, Makoto Wada, Akira Uno, Tadahito Nadamoto, Shinbo Minami and Masaya Ito are used for packaging designs and they are popular as souvenirs. CHA・CHA・CHA (800 yen) includes 5 varieties of chaba, each package is enough to brew in kyusu for one brew. This set allows customers to compare the tastes of 3 varieties of sencha, 1 kamairicha and 1 hojicha. Easy-to-use chaki, such as kyusu, are displayed and sold at the store as well.

(H) 11:00-18:00 (T) 03-3571-1211 (C) Monday (Ad) 5-5-6 Ginza, Chuo-ku (Ac) 1-minute walk from Ginza Station Exit B3 (Tokyo Metro) (U) www.uogashi-meicha.co.jp/en

㉑ 前进屋茶店（自由之丘）

前进屋茶店开在高级住宅区自由之丘，从自由之丘车站出来，沿着商业街走几步，便能看到在街角的绿色与水色外墙的门面店，茶壶图案的招牌挂在外墙上尤为显眼。这家店主要销售来自鹿儿岛的茶叶。前进屋原本是鹿儿岛一家历史悠久的茶叶批发店，2012 年在鹿儿岛开设了第一家茶店后，2016 年于东京自由之丘开设了这家分店。为了推广产自鹿儿岛的茶叶，前进屋不仅有原创的茶品牌，还会销售严选自固定茶园的茶叶。其中最值得推荐的是"kokumaro"，这款茶是将茶叶用独创的文火煎炒技法制成的混合茶，力求保留茶叶原本的味道，这样制成的茶叶有天然的甘甜，闻起来有桉树的香气。绿色的鲜艳茶叶包装分为 100 克装（1200 日元）、200 克装（2040 日元），除此之外还有 20 克装（500 日元）的试饮装。店内还出售茶壶与茶叶罐这类茶具，以及跟每款茶搭配的点心，既适合带回家享用，也可以作为送礼佳品。

㊐时 10:00—19:00 ㊐电 03-6421-4142 ㊐休 每月第一个及第三个星期三（节假日顺延）㊐址 目黑区自由之丘 1-25-5 ㊐交 东急东横线自由之丘站出口步行 3 分钟 ㊐网 susumaya.com

第四章

㉑ susumuya chaten (Jiyūgaoka)

Exit from Jiyugaoka Station, then walk a little along the commercial strip with a covered arcade, you will see a sign with a kyusu shaped logo hanging down. That is Susumuya Chaten. Susumuya Chaten is a retail store for tea wholesalers with a long business history in Kagoshima. The first store of susumuya chaten was opened in Kagoshima in 2012 and this store was opened in 2016. It sells carefully selected nihoncha, including its own blend and nihoncha from a particular tea farm, to let customers know of nihoncha products from Kagoshima. The most popular brand is Kokumaro, in which cho-asairi,very lightly roasted, almost raw nihoncha leaves are added. It has natural sweetness and a rich taste, and its aroma reminds you of eucalyptus. 100g (1,200 yen), 200g (2,040 yen) and sample package in 20g (500 yen) are available for sale. Original items, such as kyusu or chazutsu, in addition to sweets you can enjoy pairing with nihoncha are sold at the store, perfect for enjoying at home or giving them as gifts.

(H) 10:00-19:00 (T) 03-6421-4142 (C) First and Third Wednesdays (If a Holiday falls on a First or Third Wednesday, then Thursday will be closed) (Ad) 1-25-5 Jiyugaoka, Meguro-ku (Ac) 3-minute walk from Jiyugaoka Station Front Exit (Tokyu Toyoko Line) (U) susumuya.com

㉒ 椿宗善广尾店（广尾）

椿宗善广尾店的长田店主在鸟取县老家有一家家族经营的日本茶专卖店“长田茶店”，他曾将老家的茶铺开到了涩谷东急 Plaza 商场，但是由于东急 Plaza 商场于 2015 年歇业而不得已将店铺关掉。在那之后，长田成为了福井茶专卖店“椿宗善”的加盟店店主。椿宗善广尾店于 2015 年开张，店内有大约 200 种茶叶，不仅有绿茶，还提供调味茶与健康茶饮。广尾地区靠近涩谷，有较多年轻女性、情侣以及外国游客。包装鲜艳的伴手礼排列在靠近入口的位置，新茶试饮的位置在店铺中央，这种充满轻快氛围的店铺让人忍不住进去“一探究竟”。除了椿宗善主打的福井茶之外，长田本人也会亲自制作混合茶，还出售名为“女流茶人长田佳子作”的煎茶，以照顾在涩谷开店以来相熟的老顾客。今后店主还打算发掘更多的茶品，以满足客人各方面的需求。

㊋ 10:00—19:00 ㊔ 03-6455-7581 ㊡ 每月第四个星期二 ㊑ 涩谷区广尾 5-8-1 ㊍ 地铁日比谷线广尾站步行 5 分钟 ㊓ tsubakisozen.co.jp

㉒ TSUBAKI SOZEN Hiroo (Hiroo)

Ms. Nagata, the store manager of TSUBAKI SOZEN, was managing a Tokyo store of the teashop owned by her family called Nagata Chaten, which had a history since 1801. She closed the Tokyo store when Tokyu Plaza in Shibuya closed. In 2015, she opened a new store in Hiroo, partnering with TSUBAKI SOZEN, a nihoncha specialty store in Fukui. Currently, the store offers many varieties of teas, including flavored teas or health tea which are popular among young women and foreign tourists in Hiroo. This is a street-front store, so products packed in colorful packages, and gift items are displayed at the front. Customers can enjoy seasonal tea samples at the center of the store, while enjoying its open atmosphere. It also continues to offer nihoncha products of Nagata Chaten. Nagata's original sencha using gogumi, means blend, has Ms. Nagata's name and named Jyoryu Chajin, Yoshiko Nagata. She continues to cherish connections with her clients from her time in Shibuya. Café's product line-ups and clientele is expanding and it will be interesting to see how this store develops.

(H) 10:00-19:00 (T) 03-6455-7581 (C) 4th Tuesday (Ad) 5-8-1 Hiroo Shibuya-ku (Ac) 5-minute walk from Hiroo Station (Tokyo Metro Hibiya Line) (U) tsubakisozen.co.jp

㉓ Ocharaka（日本桥）

Ocharaka 是一家调味茶专卖店，店主斯蒂芬·丹顿来自法国，已在日本定居 20 年。茶店开在日本桥地区，进入店内便会被柜台后面排列整齐的茶箱吸引，扑面而来的是茶叶的清香以及花果的甜香。店内提供如“夏橙”“桃子”等 50 多种口味的调味茶。斯蒂芬初次在日本喝到日本茶时，便开始考虑如何将日本茶打造成一种“软饮料”。为了让不喜欢喝茶的人也能接受并对茶感兴趣，他着手制作各类调味茶，比如在茶叶中加入花瓣，让茶叶看起来色彩更丰富，从而更好地品尝日本茶原本的香味。这位法国专业品酒师，将他品鉴红酒的专业知识有效地结合在了调味茶的调配上，他调配的茶会给人带来视觉、嗅觉以及味觉上的三重享受。Ocharaka 于 2005 年在吉祥寺开业，于 2014 年搬迁到日本桥，店里经常会有年轻人与外国游客光顾。斯蒂芬还会不定期地举办试茶会以及迷你演讲，以传播日本茶的魅力。店内还销售其他各类茶品。

㊡ 10:00—21:00 ㊥ 03-6262-1505 ㊡ 根据 COREDO 室町商场的营业时间 ㊟ 中央区日本桥室町 2-2-1COREDO 室町商场 1-B1 层 ㊣ 地铁银座线三越前站 A6 出口 ㊫ www.ocharaka.co.jp

㉓ OCHARAKA (Nihonbashi)

Inside the store where eye-catching wooden boxes are lined up, your nose will be tickled by the sweet aroma and tea aroma. OCHARAKA is a specialty store of flavored tea, fruits or flower flavored ryokucha or hojicha. Its flavored teas come in about 50 varieties, such as Natsumikan (summer orange) or Momo (peach). The store manager, Stefan Danton, came to Japan 20 years ago, and saw nihoncha's possibility as a soft drink. He sought a broader acceptance of nihoncha by developing flavored nihoncha. Even people who are not familiar with nihoncha can enjoy the visual beauty of added flowers, the appetizing aroma, and its fundamental deep flavors. Stefan has a Sommelier Certificate-guests can enjoy nihoncha with their "eyes・nose・mouth" as if they are enjoying wine. OCHARAKA provides tasting and mini-seminars. After the café moved to the current location in Nihonbashi, the store started offer tasting and mini-seminars for foreigners, to teach them about fascinations of nihoncha. At OCHARAKA, nihoncha from Japan is also sold.

(H) 10:00-21:00 (T) 03-6262-1505 (C) Same hours as COREDO Muromachi's operation hours (Ad) COREDO Muromachi 1-B1, 2-2-1 Muromachi, Nihonbashi, Chuo-ku (Ac) Near Mitsukoshi-Mae Station, Exit A6 (Tokyo Metro Ginza Line) (U) www.ocharaka.co.jp/english

やきいも
マロングラッセ
バースデイ
スパイシー
バナナチョコ

在日本桥店开张的时候，斯蒂芬最看重的是背后这座可以放入所有茶箱的落地柜，以及面前这个全白的长形柜台。在这里他有足够的空间将多个种类的茶叶排列开来，进行试茶对比。

At the new store, Stefan paid special attention to the shelf where he can line up all wooden tea boxes for all flavored teas, as well as the white counter, where he can spread chaba or conduct tastings.

店内多处装潢都采用桐木、土以及和纸等和风材料，效仿旧式茶屋以及日式仓库的装修风格，这样的空间让人心情莫名地平静。

At the store, items typical to Japanese buildings, such as wood, earth and papers are used. They provide a nostalgic atmosphere to the store, reminding customers of old, Japanese ochayasan (teashop) or kura (Japanese stone house).

调味茶的主要原料是日本各地产的绿茶，店主认为通过比较茶叶的呈色、形状以及香气，就能清楚了解每种茶的特征。店内会不定期地举办试茶会，让参加者亲自去对比各种茶叶的不同。

When you compare the appearance and aromas of nihoncha, used as a base for flavored tea, from all over Japan, you will understand the depth of nihoncha. Tasting events in which guests can compare several varieties of nihoncha are held at the store on irregular schedules.

Ocharaka 以日本茶为基础的调味茶的魅力

FASCINATION OF THE FLAVORED TEA WITH NIHONCHA USED AS A BASE TEA

这款“室町”茶是为庆祝日本桥店开业时配制的，色香味上都带着祝福的意愿。茶的具体配方是在带有香槟香味的绿茶中，混入粉玫瑰、木槿花等干花，再加入一些制作蛋糕时会用上的银色糖粒。茶叶冷萃过后，泡出的茶水呈现一种清透的金色，宛如香槟酒一般。Ocharaka 通过这样的创新搭配，不仅保留了优质日本茶的原味，还拓展了日本茶的喝法以及调味茶在制法上的无限可能性。

Muromachi was a blend created for the opening of the Nihonbashi store.Champagne flavored ryokucha with dragée and flowers, such as pink rose and hibiscus added. Muromachi's looks and aroma are perfect for celebrations.When cold brewed with water, it looks as if it is a glass of champagne with its golden, clear color.The possibilities of flavored teas, created based on the high-quality nihoncha, are unlimited.

其他推荐
BEST CAFES and MORE

三木园（饭田桥）

开业至今 70 年，专注于销售自家以及合作茶园的茶叶。

(时) 8:30—19:00（星期六 8:30—17:00）(电) 03-3234-6819 (休) 星期日、节假日、每月第三个星期六 (址) 千代田区饭田桥 1-9-7 (交) JR 饭田桥站东出口步行 5 分钟或东西线饭田桥站 A5 出口步行 1 分钟 (网) www.e-mitsugien.co.jp

Mitsugien (Iidabashi)

70 years since opening. Only serves nihoncha from own chaen or contracted farms

(H) 8:30-19:00 (Saturday: 8:30-17:00) (T) 03-3234-6819 (C) Sunday, Holidays and Third Saturday (Ad) 1-9-7 Iidabashi, Chiyoda-ku (Ac) 5-minute walk from Iidabashi Station East Exit (JR Line):1-minute walk from Iidabashi Station Exit A5 (Tokyo Metro Tozai Line) (U) www.e-mitsugien.co.jp

---

下北茶苑大山（下北泽）

日本茶专卖店，店内有两位拥有十段茶师资质的店员。

(时) 10:00—20:00（星期三 10:00—18:00）(电) 03-3466-5588 (休) 元旦 (址) 世田谷区北泽 2-30-2 (交) 小田急线或京王井之头线下北泽站北口步行 2 分钟 (网) shimokita-chaen.com

Shimokita Chaen Oyama (Shimokitazawa)

Nihoncha specialty store with 2 level 10 dan Chashi - tea master.

(H) 10:00-20:00 (Wednesday 10:00-18:00) (T) 03-3466-5588 (C) New Year's Day (Ad) 2-30-2 Kitazawa, Setagaya-ku (Ac) 2-minute walk from Shimokitazawa Station North Exit (Odakyu Line and Keio Inokashira Line) (U) shimokita-chaen.com

---

坪市制茶总店浅草店（浅草）

大阪界市的老牌茶叶铺东京分店。

(时) 10:00—20:00 (电) 03-3841-0155 (休) 无 (址) 台东区浅草 2-6-7 Marugoto 日本 2 楼 (交) 筑波快线浅草站步行 2 分钟 (网) www.tsuboichi.co.jp/saryo

TEA TSUBOICHI the Tea House (Asakusa)

Retail store & café of an established nihoncha specialty store in Sakai, Osaka.

(H) 10:00-20:00 (T) 03-3841-0155 (C) None (Ad) Marugoto Nippon 2nd Floor, 2-6-7 Asakusa, Taito-ku (Ac) 2-minute walk from Asakusa Station (Tsukuba Express) (U) www.tsuboichi.co.jp/saryo

---

大桥（中野）

建筑设计和室内装潢带有复古风格的茶店。

(时) 10:30—19:00 (电) 03-3381-5320 (休) 星期日 (址) 中野区中野 3-34-31 (交) JR 中央线或地铁东西线中野站南口步行 2 分钟 (网) ohashi-cha.blogspot.jp

Ohashi (Nakano)

The building design and interiors create an atmosphere as if the store is an antique shop.

(H) 10:30-19:00 (T) 03-3381-5320 (C) Sunday (Ad) 3-34-31 Nakano, Nakano-ku (Ac) 2-minute walk from Nakano Station South Exit (JR Chuo Line and Tokyo Metro Tozai Line) (U) ohashi-cha.blogspot.jp

从东京到静冈，参观日本茶工厂

TAKE A TRIP FROM TOKYO, NIHONCHA FACTORY TOUR IN SHIZUOKA

伊藤园是日本有名的茶饮品牌。它在静冈县牧之原市的茶叶加工工厂——静冈相良工厂向一般民众开放参观，可参观制作茶包的生产线，也可以在工厂的“中央研究所”内学习茶叶生产线的质检管理体制。每周二 13：30 以后可以入厂参观，接受个人申请，申请免费，需要提前预约。

(时) 9:00—16:00 (电) 0548-54-0311 (休) 星期六、星期日、节假日 (址) 静冈县牧之原市女神 21 (交) 在 JR 金谷站搭乘出租车行驶约 30 分钟至东名高速路相良牧之原出口后，在 473 号国道南下行驶约 20 分钟

Itoen is known for its nihoncha. Makinohara is famous for its high-quality tea and Itoen's Sagara Factory is located here. Tour includes the packaging line and the Research Center attached to the factory. Individual tours are held every Tuesday from 13:30. Free of charge, but advanced reservation required.

(H) 9:00-16:00 (Hours of Inquiry) (T) 0548-54-0311 (C) Saturday, Sunday and Holiday (Ad) 21 Mekami Makinohara, Shizuoka (Ac) 30-minute on Taxi from Kanaya Station (JR Line), Exit Tomei Highway Sagara Makinohara IC, Go toward South on Japan National Route 473, 20-minute drive

在前台登记后，会有工作人员带领依次参观工厂及“中央研究所”。在工厂主要是参观包装生产线是如何运作的；在研究所，会有工作人员讲解伊藤园为了保证茶饮的美味以及生产线的安全，分别做了哪些研究与检测。

After registration, you will be taken on a tour from the factory first. At the factory, you will observe the packaging line. At the Central Research Center, you will learn about what research and tests are conducted in order to maintain the taste and safety of the products.

小百科 / GLOSSARY

薄茶：日本茶道茶会中冲泡抹茶时可分为薄茶与浓茶两个浓度，且出现在不同的茶会环节当中。

点茶：茶道茶会中的一个环节，指抹茶制作的整个过程，包括舀水、冲水、打泡等多个步骤，各个流派的点茶方式也不同。

和三盆：日本四国地区东部生产的传统砂糖。

茎茶：在玉露和煎茶的制作过程中，挑选新芽上的茶茎加以烘焙、炒制而成。

蒸煎茶：煎茶过程中，茶叶的热处理时间比较长。成品茶形态较碎、颜色较深、茶味比较浓厚。

蹲踞：日本庭园中常常会设置一个洗手池，在传统茶会中，茶客进入茶室前都需要在庭院中等待，并在“蹲踞”处洗手后才可以进入茶室。

茶之间：家庭成员聚在一起吃饭、休闲的起居室。

番茶：绿茶的一种，茶叶品质较低，多采用夏天后采收的绿茶茶叶制成。

碾茶：抹茶粉的原料，指还没被磨成粉的茶叶。

NOTES

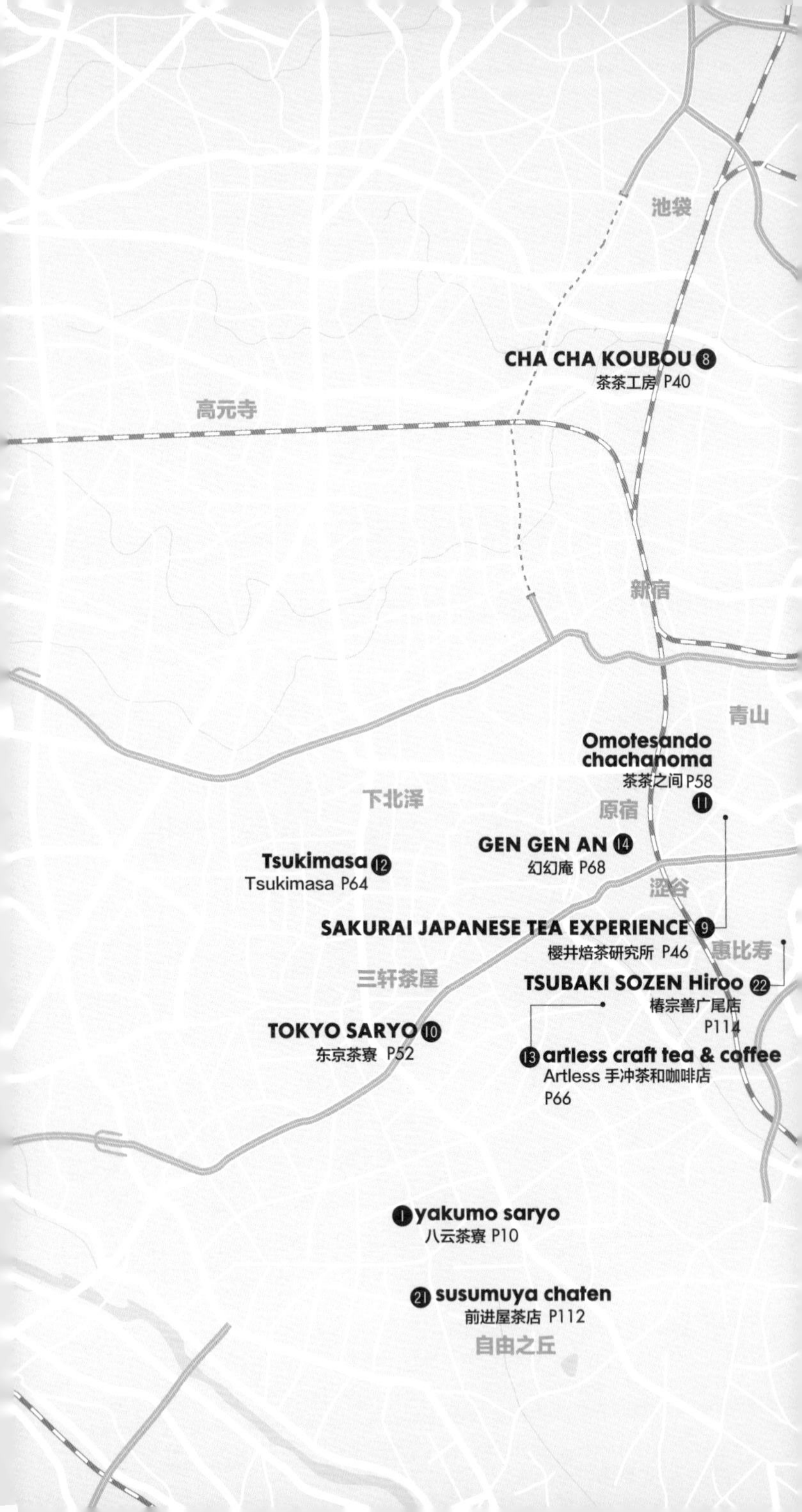

池袋
CHA CHA KOUBOU 8
茶茶工房 P40
高元寺
新宿
青山
Omotesando chachanoma
茶茶之间 P58
11
下北泽
原宿
GEN GEN AN 14
幻幻庵 P68
Tsukimasa 12
Tsukimasa P64
涩谷
SAKURAI JAPANESE TEA EXPERIENCE 9
樱井焙茶研究所 P46
惠比寿
三轩茶屋
TSUBAKI SOZEN Hiroo 22
椿宗善广尾店
P114
TOKYO SARYO 10
东京茶寮 P52
13 artless craft tea & coffee
Artless 手冲茶和咖啡店
P66
1 yakumo saryo
八云茶寮 P10
21 susumuya chaten
前进屋茶店 P112
自由之丘

6 Gekko
月光 P36
上野
浅草
东京天空树
Rakuzan
乐山 P104
17
2 AKANE-YA
茜夜 P18
秋叶原
16 NAKAMURA TEA LIFE STORE
中村茶生活馆 P98
乐坂
两国
23 OCHARAKA
Ocharaka P116
皇宫
丸之内
5 Yamamotoyama Nihonbashi Honten
山本山日本桥本店茶室 P34
IPPODO TEA
okyo Marunouchi
Store
15
一保堂茶铺东京丸之内店
P74
4 HIGASHIYA GINZA
东屋茶寮 P28
银座
19 ISHIZAKIEN
石崎园 P108
7 Jugetsudo Ginza Kabukiza
寿月堂银座歌舞伎座店 P38
六本木
东京塔
20 Uogashi-Meicha CHA·GINZA
鱼河岸茗茶银座店 P110
3 NAKAMURA TOKICHI HONTEN Ginza Store
中村藤吉本店银座店 P24
18 TSUJIRI Ginza
辻利银座店 P106
品川

日文版工作人员

Cover Illustration: NORITAKE　Designer: TUESDAY (Tomohiro+Chiyo Togawa)
Map: Manami Yamamoto (DIG.Factory)
Photographer: Yosuke Otomo (except P34-35 , 38-39 , 66-67 , 96-97, 106-107 , 112-113 , 122)
Japanese Writer: Miho Ohara　Translator: Kyoko Shakagori
Proofreader (Japanese): Terumi Arakawa　Proofreader (English): Jonathan Berry
Editorial Director: Miki Usui (BIJUTSU SHUPPAN-SHA CO., LTD.)

图书在版编目（CIP）数据

日本茶 / 日本美术出版社书籍编辑部编著 ; 黄迪译
. -- 北京 : 中信出版社 , 2019.7（2019.9 重印）
（东京艺术之旅）
ISBN 978-7-5217-0422-8

Ⅰ. ①日… Ⅱ. ①日… ②黄… Ⅲ. ①茶文化－介绍
－日本 Ⅳ. ① TS971.21

中国版本图书馆 CIP 数据核字 (2019) 第 073294 号

日本茶

编　　著：【日】美术出版社书籍编辑部
译　　者：黄迪
出版发行：中信出版集团股份有限公司
（北京市朝阳区惠新东街甲4号富盛大厦2座　邮编　100029）
承 印 者：北京雅昌艺术印刷有限公司

开　　本：880mm × 1230mm　1/32　　印　　张：4
字　　数：60千字　　版　　次：2019年7月第1版
印　　次：2019年9月第4次印刷　　京权图字：01-2019-2331
广告经营许可证：京朝工商广字第8087号
书　　号：ISBN 978-7-5217-0422-8
定　　价：45.00元